Bioavailability, Leachability, Chemical Speciation, and Bioremediation of Heavy Metals in the Process of Composting

Bioavailability, Leachability, Chemical Speciation, and Bioremediation of Heavy Metals in the Process of Composting

Jiwan Singh and Ajay Kalamdhad

CRC Press is an imprint of the
Taylor & Francis Group, an informa business

CRC Press
Taylor & Francis Group
6000 Broken Sound Parkway NW, Suite 300
Boca Raton, FL 33487-2742

Printed on acid-free paper

International Standard Book Number-13: 978-1-138-59833-1 (Hardback)

Library of Congress Cataloging-in-Publication Data

Names: Singh, Jiwan (Professor of environmental science), author. |
Kalamdhad, Ajay, author.
Title: Bioavailability, leachability, chemical speciation, and bioremediation
of heavy metals in the process of composting / Jiwan Singh, Ajay Kalamdhad.
Description: Boca Raton: CRC Press, Taylor & Francis Group, 2019. | Includes
bibliographical references and index.
Identifiers: LCCN 2018014317 | ISBN 9781138598331 (hardback) |
ISBN 9780429486395 (ebook)
Subjects: LCSH: Compost. | Heavy metals. | Bioremediation. | Speciation
(Chemistry)
Classification: LCC TD796.5 .S56 2019 | DDC 631.8/75--dc23
LC record available at https://lccn.loc.gov/2018014317

Visit the Taylor & Francis Web site at
http://www.taylorandfrancis.com

and the CRC Press Web site at
http://www.crcpress.com

Contents

List of Figures

List of Figures

List of Tables

Preface

The treatment and disposal of biodegradable solid wastes, predominantly wastes generated in small, medium, and large urban cities, have become a serious problem mainly in developing countries. In the developing countries, the composting of organic waste and applied compost to land signifies one of the best methods to treat/dispose of biodegradable solid waste since it solves two problems: one is material recycling and the other is to manage the solid waste disposal problem. Composting is the best-known process for the stabilization of waste by converting it into more stabilized and pathogen-free material (compost) which can be applied in agricultural fields. If high contents of toxic heavy metals are present in the final compost to be applied for agricultural lands, the absorption of these toxic heavy metals by growing plants and the succeeding accumulation in the food chain may cause a potential risk to animal and human health.

The book is organized into eight chapters. These chapters cover different types of composting processes and physicochemical changes occurring during the composting process. A further chapter is devoted to the method of sampling and analysis of compost samples that can be used for both quantitative and qualitative analyses.

Chapter 1 consists of a brief discussion of different types of waste materials; treatment through different composting techniques; heavy metals occurring in the final composting; effects of heavy metals on the environment; bioavailability, leachability, and speciation of heavy metals; and metal reduction approaches during the different composting process. This chapter explains the impact of heavy metals on soil quality, aquatic life, plants, and human beings. This chapter also discusses the effects of metals on the environment.

Chapter 2 deals with the composting process, factors affecting the composting process, types of composting processes, heavy metals in the composting biomass, and the effects of metals on composting process.

Chapter 3 provides a detailed analysis of compost samples and heavy-metal bioavailability such as water solubility, DTPA extractability, and chemical speciation. This chapter also deals with the application of the physicochemical and biological parameters.

Chapter 4 discusses the effect of heavy metals on the composting process.

Chapter 5 evaluates the effect of the bioavailability of heavy metals during the composting process. The bioavailability of metals includes water solubility and extractability of metals using diethylene diamine penta-acetic acid (DTPA).

Chapter 6 explains the different forms of heavy metals such the exchangeable, carbonate, reducible, oxidized, and residual fractions. This chapter also deals with a detailed speciation study of Cd, Cr, Cu, Fe, Mn, Ni, Pb, and Zn in the composting of various wastes.

Chapter 7 evaluates the effects of chemical addition (lime, sodium sulfide, red mud, and natural zeolite) on the bioavailability and speciation of heavy metals during the composting of different wastes. This chapter also explains the role of alkaline chemicals for the reduction of the bioavailability.

Chapter 8 explains the role of microorganisms for the reduction of bioavailability of heavy metals in the composting process.

Chapter 9 explains the leachability of heavy metals in the composting process of different wastes. This chapter also evaluates the effects of different chemicals, such as lime, natural zeolite, sodium sulfide, etc., on the leachability of heavy metals.

This publication has been prepared primarily to serve two audiences: (i) researchers working in the field of composting of different types of solid wastes including industrial wastes and (ii) professionals involved in the management of solid wastes. The information presented in this book would also be useful for students of environmental science and engineering.

AIM AND SCOPE OF BOOK

The aim of this book is to support all those involved in research and teaching related to composting and heavy-metal study in the leading academic and research organizations around the world. This book will be of great value to postgraduate students, research scientists, and professionals in environmental science and engineering, soil science, environmental health, and authorities responsible for the management of metal-contaminated wastes.

Acronyms and Abbreviations

F1	exchangeable fraction
F2	carbonate fraction
F3	reducible fraction
F4	fraction bound to organic matter
F5	residual fraction
APC	agitated pile composting
BV	bamboo vinegar
C/N	carbon/nitrogen
EC	electrical conductivity
GW	green waste
MC	moisture content
MSW	municipal solid waste management
OM	organic matter
PM	pig manure
RDC	rotary drum composting
SS	sewage sludge
TKN	total Kjeldahl nitrogen
TN	total nitrogen
TP	total phosphorus

Authors

Dr. Jiwan Singh is working as an assistant professor in the Department of Environmental Science, Babasaheb Bhimrao Ambedkar University (A Central University), Lucknow, India since 2016. He received MS in Environmental Science from the Babasaheb Bhimrao Ambedkar University, and a Master of Technology degree in Environmental Engineering from National Institute of Foundry and Forge Technology, Ranchi, India and a PhD in Civil Engineering from the Indian Institute of Technology Guwahati (IITG), India. He worked as a post-doctoral research fellow at the Department of Civil and Environmental Engineering, University of Ulsan, South Korea, from 2013 to 2015. He also worked as an assistant professor at Department of Environmental Engineering, Kwangwoon University, Seoul, South Korea, from March, 2015 to June 24, 2016. He qualified Graduate Aptitude Test of Engineering (GATE) in 2008 and University Grant Commission (UGC) National Eligibility Test (NET) in 2013. He received UGC-BSR research start up grant project from UGC, New Delhi and early career research award from the Department of Science and Technology, Govt. of India. He published more than 55 research papers and book chapters. His research and teaching interest are in Environmental Science and Engineering including solid waste management, water and wastewater treatment, recovery of resources from the solid wastes and Environmental Nanotechnology.

Dr. Ajay Kalamdhad is working as associate professor in the Department of Civil Engineering, Indian Institute of Technology (IIT) Guwahati obtained his Bachelor (2001), Masters (2003), and Doctorate (2008) degrees in Civil and Environmental Engineering from GEC Jabalpur, VNIT Nagpur, and IIT Roorkee, respectively. Prior to joining IIT Guwahati in 2009, He worked as assistant professor at VNIT Nagpur (2008–2009) and worked in various projects at RRL, Bhopal (Now AMPRI, Bhopal), and NEERI, Nagpur. He has published more than 110 international papers in peer-reviewed journals and presented his work in more than 200 national and international conferences/workshops. He has also associated with Indian Public Health Engineers, India, International Solid Waste Association Italy, National Solid Waste Association of India, and with reviewers of 50 international journals. He is a recipient of ISTE- GSITS national award for best research by young teachers (below 35 years) of engineering colleges for the year 2012 and IEI Young Engineers Award 2011–2012 in Environmental Engineering discipline from Institute of Engineers India. He had worked on many major research projects with MoEF, DST, MDW&S, and CSIR, Govt. of India.

Authors

1 Introduction

1.1 OVERVIEW

The rapid rate of urbanization and industrialization has led to an increase in municipal solid waste (MSW) worldwide. Sewage sludge (SS), which is generally produced from wastewater treatment plants, plays a vital role in generating MSW. SS mainly comprises organic materials, macro/micronutrients, nonessential trace elements, microorganisms, and eggs of parasitic microorganisms (Wang et al., 2008). Nonessential trace elements decrease crop production and threaten soil safety and human health. A disposal and environmental management of nonessential trace elements are a worldwide concern (Wang et al., 2013). Water hyacinth (*Eichhornia crassipes*) is considered a green waste (GW) for its removal of organic and inorganic pollutants from water and soil. Water hyacinth is one of the most commonly used aquatic plants in constructed wetlands due to its fast growth rate and ability to absorb nutrients and contaminants (Singh and Kalamdhad, 2012). Composted residuals derived from MSW and GW have a high affinity to form organic complex of metals. There is a consensus in scientific literature that composting processes increase the complexation of heavy metals in organic waste residuals. Heavy metals are strongly bound to the organic fraction of organic matter (OM) and metal complexes, thus limiting their solubility and potential bioavailability in the soil (Smith, 2009).

Composting is the biological decomposition and stabilization of organic substrates (SS, GW, etc.), under conditions that allow the development of thermophilic temperatures as a result of biologically produced heat, to produce a final product that is stable, free of pathogens and plant seeds, and can be successfully applied to agricultural land (Singh and Kalamdhad, 2013a).

Aerobic microorganisms play a significant role in the decomposition of organic wastes in the composting process. Composting proceeds under three major phases: (1) mesophilic stage, (2) thermophilic stage, and (3) cooling (Neklyudov et al., 2008). Composting is a well-established technique for the treatment and disposal of MSW and GW due to material recycling (Villasenor et al., 2011). The process of composting involves the decomposition and degradation of organic waste by converting it into a more stabilized, humus-like material which is called compost. When applied on agricultural lands, it can significantly improve the physical properties and agricultural productivity of soils (Deka et al., 2011; Gabhane et al., 2012). Substances that are not biodegradable and have high concentrations of heavy metals do not decompose in the solid and hence cannot be used to improve the soil fertility of agricultural lands.

Heavy metals can accumulate in plants, which are then transferred into the food chain, and thus may cause a possible risk to animals and human beings (Iwegbue et al., 2007; Singh and Kalamdhad, 2012). The presence of heavy metals in compost can affect the growth, morphology, and metabolism of the soil microorganisms, thus

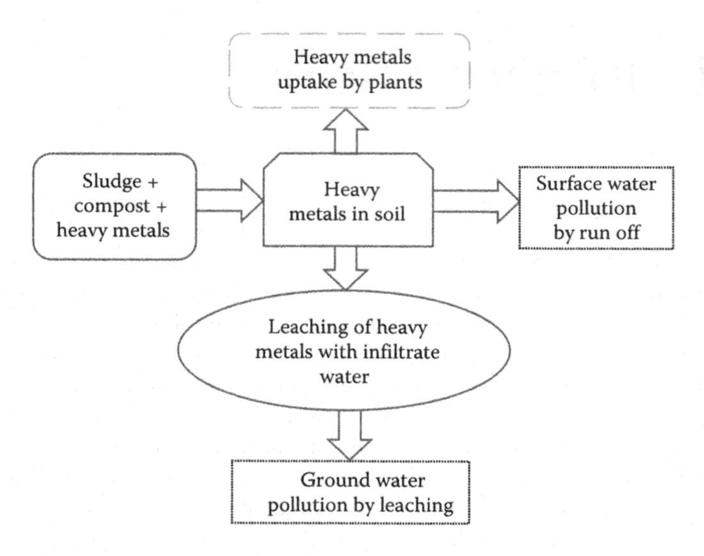

FIGURE 1.1 Fate and transport of heavy metals in the environment.

decreasing the activity of microorganisms, and soil fertility (Bragato et al., 1998). The absorption of heavy metals by plants and the corresponding accumulation in human tissue and biomagnifications through the food chain not only causes environmental concerns but also human health concerns (Wong and Selvam, 2006). Figure 1.1 represents the fate and transference of heavy metals from compost to soil, plants, and the water system.

Thus, heavy metals are considered one of the major sources of soil pollution. Though, presence in various fractions of metals and their association with ecotoxicity is the major problem. Heavy metals are natural components of the environment and have high atomic weight with a density of >4 g/cm^3 (Tchounwou et al., 2012). These metals are referred to as *trace elements* at a concentration level of <10 ppm in the environment (Tchounwou et al., 2012). Some heavy metals (such as Cu, Co, Fe, Mn, and Zn), are required for physiochemical and biochemical functions of both plants and animals. On the other hand, excessive additions of these heavy metals to the environment may have an adverse effect on both the animals and humans. Other nonessential metals (such as, Cd, Pb, Hg, and Ni), which are involved in biological functions and are serious to metal-sensitive enzymes, can initiate the cell death (Tchounwou et al., 2012; Swati and Hait, 2017). Determination of total concentration of heavy metal in compost/soil is indicator of pollution in soil but providing information about bioavailable fractions of metals. (Singh and Kalamdhad, 2012, 2013a). The bioavailability and speciation of heavy metals during the composting process suggested that toxicity of heavy metals depends on their different bioavailable fractions rather than total metal concentration detected in the digested solution (Kang et al., 2011).

The bioavailability of metals is used to indicate that part of the concentration of the metals are easily soluble in water and considered effortlessly accessible to plants and soil microorganisms. The bioavailability of metals in soil is a self-motivated process that depends on the explicit physicochemical parameters of soil. These parameters, such as pH, OM content, redox potential, cation exchange capacity of soil, and hydroxides, are present in soil, soil texture, and clay content (Guala et al., 2010). The pH and OM contents are the most important parameters involved in the absorption of heavy-metal by living organisms (Li et al., 2010).

The addition of soluble organo-metal complexes occurring in the final compost in the soil increases the potential risk of metals, since soluble metals will be more available to receptors (plants, microorganisms, etc.) (Zheng et al., 2007).

The water soluble fraction of metal is most readily bioavailable in compost when applied to soils. Hsu and Lo (2001) reported that water-soluble metals play a critical role in maximizing the potential for contamination of the food chain and water system (surface water as well as ground water).

It has been considered that the metals extracted with diethylene triamine pentaacetic acid (DTPA) solution may play a significant role in checking the bioavailability of heavy metals in soil and compost-added soil available for plant uptake (Fuentes et al., 2006; Singh and Kalamdhad, 2013b).

The toxicity characteristic leaching procedure (TCLP) test has been applied to assess the mobility of hazardous metals that are present in the compost/waste materials. If heavy metals contaminated compost is applied to soil, metals can be leached out in soil and may pollute ground water (Singh and Kalamdhad, 2013b). Leaching of metals is defined as the ratio of the amount of a heavy-metal discharge from the TCLP test to its total concentration; it is usually applied to evaluate the potential leaching ability of heavy metals occurring in the compost and amended soil (Chiang et al., 2007).

An estimation of the fractionation of heavy metals in the final compost helps to assess their bioavailability and the aptness of compost to be applied for land application (Wong and Selvam, 2006). The sequential extraction method affords a useful technique for determining the different chemical forms of heavy metals in sludge/compost. This technique has been applied to organic soils, which are very similar to composts, for studying metal mobility and bioavailability (Yuan et al., 2011). The mobility of the heavy metals is decreased roughly in the order of extraction categorization (Nair et al., 2008). Tessier et al. (1979) reported a sequential extraction method of heavy metals that are present in the compost. These metals involve the following five fractions: (i) exchangeable fraction (F1): this fraction can be transformed by changing ionic composition of water in addition to sorption–desorption process, (ii) carbonate fraction (F2): this fraction is depend on pH and have ability to convert into soluble and mobilized fractions under acidic condition, (iii) reducible fraction (F3): this fraction is thermally disturbed under anoxic conditions, (iv) oxidizable fraction (F4): this fraction will be released and solubilize after getting oxidizing conditions, and (v) residual fraction (F5): this fraction is perpetually bound with the mineral components of the compost and soil. This fraction will never be accessible for plants or soil microorganisms in standard natural conditions. In the sequential extraction procedure various chemical reagents are applied to extract various form of metals from the compost sample, chemical reagents are elaborate in dissolving the different constituents of the sample medium in order sequentially. On the contrary, a reagent should have the ability to liberate all fractions of the metals from a specific constituent of matrix (i.e., F1 and F2 fractions) without touching other constituents (Li et al., 2001). The sequential extraction technique provide important information about different fractions of metals and allows the forecast of metal leaching rates (He et al., 2009).

Some of the major sources to decrease metal noxiousness for the environment including human health are reduction of heavy-metal bioavailability, leachability, and their various fractions by the addition of a few chemicals while composting different waste materials. Lime is an alkaline material and can lead to a decrease in the mobility of heavy metals in the final compost (Chiang et al., 2007). Adding lime in composting biomass increases its pH level, thus resulting in a decrease in the accumulation of metals in soil (Wong and Fang, 2000). Lime can be used as a stabilizer for heavy metals that will increase the rate of degradation of organic materials in the composting process by offering a buffering to composting biomass during acid formation after degradation of organic materials resulting decrease in pH. Lime addition in appropriate supply sufficient amount of Ca to the composting microbes that enhanced the metabolic activity of microbes resulting in an increase in temperature and CO_2 progress without any negative impact on microbial community present in the composting mass (Fang and Wong, 1999; Gabhane et al., 2012). The addition of lime was very efficient in reducing bioavailability of heavy metals in the mature compost of SS. It might form less-soluble carbonate salts with metals ions (Fang and Wong, 1999). Zeolite is a naturally hydrated aluminosilicate mineral and can be classified as "tectosilicate" (Singh and Klamdhad, 2014). Zeolites can be used extensively in the composting process to improve physical and chemical characteristics of the compost and then immobilize metals in the SS composting (Sprynskyy et al., 2007; Villasenor et al., 2011). It improves the composting process by increasing the porosity of the substrate of the composting mixture (Zorpas et al., 2000). It has the ability to readily absorb almost all heavy metals that are bound to exchangeable and carbonate fractions (Zorpas et al., 2000). It can increase sodium and potassium in the compost through exchange with toxic metals.

1.2 EFFECTS OF HEAVY METALS ON THE ENVIRONMENT

1.2.1 Effects of Metals on Soil

In recent years, natural anthropogenic activities can cause elevated concentrations of heavy metals in the soil. Consequently, the environment is deteriorating because of the negative impact of heavy metals (Su et al., 2014; Zojaji et al., 2014). The quality of the soil may be represented by the microbial and enzymatic activity of the soil. Microorganisms present in soil are an important sign of the level of pollution in the soil. Microbial activity can be prevented in the soil that has been polluted by heavy metals (Su et al., 2014). Soil pollution by heavy metals is one of most significant concerns for the whole industrialized world (Hinojosa et al., 2004).

Metals present in soil not only causes an adverse effect on different properties of soil related to plant life but also can cause deviations in composition, size, and activity of the microbial community (Yao et al., 2003). Thus, heavy metals are considered to be a main factor responsible for soil contamination. There are several metals, such as Cu, Ni, Cd, Zn, Cr, and Pb, which are responsible for soil contamination (Hinojosa et al., 2004). The soil's biological and biochemical properties are affected by heavy metals. According to Speira et al. (1999), the soil parameters, such as OM, clay contents, and pH value, have key effects on biochemical properties.

Soil enzymatic activities were affected by toxic metals indirectly through shifts in the microbial community (Shun-hong et al., 2009). A key microbial process can also be affected by heavy metals in soil by decreasing the number of soil microorganisms and their activities. If the long-term exposure of heavy metals affects the tolerance of the bacterial community and fungi (arbuscularmycorrhizal fungi), then these microorganisms can play a vital role in the restoration of metal-polluted ecosystems (Mora et al., 2005). Heavy metals can decrease the productivity of many bacterial species, can increase actinomycetes in soil, and can affect both the biomass and diversity of the bacterial communities in polluted soils (Chen et al., 2010).

Enzymatic activities of soil are affected in different ways by different metals due to the different chemical binding properties of the enzymes in the soil environment. It has been considered that Cd is highly toxic to enzymes due to its better mobility and low-binding properties when compared with soil colloids. Cd contamination had a negative effect on the activities of protease, urease, alkaline phosphatase, and arylsulfatase, whereas it did not show any significant effect on the invertase (Karaca et al., 2010). Cu toxicity inhibited the β-glucosidase activity, whereas it did not affect the cellulose activity. The Pb contamination had a negative effect on the activities of urease, catalase, invertase, and acid phosphatase (Karaca et al., 2010; Singh and Kalamdhad, 2011). Soil microbes have important roles in recycling nutrients for plant, conservation of soil structure, decontamination of harmful chemicals, plant pest control, and control of plant growth communities that represent directories of soil quality. Repeating contamination of soil by heavy metals is an important study about the functioning of soil microorganisms (Wang et al., 2007).

Chromium is commonly present in soils as Cr (III) and Cr (VI). Cr (III) is a micronutrient and a nonhazardous species, whereas Cr (VI) is a strong oxidizing agent and is 10–100 times more toxic than Cr (III) (Garnier et al., 2006). With higher concentrations, Cr (VI) could cause shifts in the composition of soil microbial populations, and is also known to cause harmful effects on microbial cell metabolism at higher level (Shun-hong et al., 2009). Ashraf and Ali (2007) reported that the toxic effect of heavy metals on soil microorganisms prompts a change in the diversity, population size, and overall activity of the soil microbial communities. Generally, an increase of metal content shows an adverse effect on soil microbial metabolic activities, such as respiration rate and enzyme activity. These activities are very valuable indicators of soil contamination. The soil microbial profile was changed with the toxicity of lead (Pb) in the soil system.

1.2.2 Effects of Metals on Plants

The availability of both essential and nonessential heavy metals at certain concentration levels in the soil can inhibit the growth of most plants. However, if the high concentration of metals is supplied to the plant that is exceeding threshold limit, resulting death of the plant (Su et al., 2014). Some heavy metals, such as As, Cd, Hg,

Pb, and Se, are not necessary elements for plant growth due to their absence in the physiological metabolism of plants, whereas some trace metals, such as Zn, Cu, Mn, Fe, Co, Mo, and Ni, are essential elements for the usual growth and metabolism of plants. However, these elements can be toxic to plants if their concentration within the soil is too high (Garrido et al., 2002; Rascio and Izzo, 2011). An application of compost to agricultural field to increase productivity of crops without concerning potential negative impact on crops, however the compost prepared from metal contaminated waste is applied to soil for growing vegetables, metals may transfer of from soil to vegetable plants, and plant to humans through edible part of plant should be considerable issue (Jordao et al., 2006). Qin et al. (1994) reported that the Cd concentration of 30 μmol/L in soils affects the growth of cabbage and bean seedling. Cd can disrupt the metabolic activities such as photosynthesis, protein synthesis of plants, membrane damage, etc. (Su et al., 2014).

The transfer of metals from soil to plants followed by their accumulation in the food chain is a potential threat to the environment and humans (Sprynskyy et al., 2007). The absorption of metals by plant roots is the main way toxic metals enter the food chain (Jordao et al., 2006). Sharma et al. (2007) reported that there are many factors that are involved in the absorption and accumulation of heavy metals in plant tissue, such as temperature, moisture, organic fraction, pH, and the nutrient level of the soil. The uptake and accumulation of Cd, Zn, Cr, and Mn in *Beta vulgaris* (Spinach) were higher during the summer season, whereas Cu, Ni, and Pb accumulated more during the winter season (Sharma et al., 2007). The higher accumulation of Cd, Zn, Cr, and Mn during the summer season may be due to the high decomposition rate of OM, which is likely to release these heavy metals in soil, in addition to high transpiration rates during the summer season (Sharma et al., 2007). Khan et al. (2008) reported that the absorption and accumulation of heavy metals in plants depend on the plant species and the efficiency of different plants in absorbing metals (Khan et al., 2008).

Bhattacharyya et al. (2008) reported that elevated concentration of Pb in soils may decrease soil productivity, and a very low concentration of Pb may inhibit some vital plant processes, such as photosynthesis, mitosis, water absorption, etc. Guala et al. (2010) reported that heavy metals are possibly toxic to plants and may cause phytotoxicity leading to many plant diseases such as chlorosis, weak plant growth, reduced nutrient uptake, and reduction in the fixation of molecular nitrogen in leguminous plants. Ashraf and Ali (2007) reported that elevated concentration of Pb in soils can cause delayed seed germination. Cu (2015) reported that Cu, Pb, and Zn have an incredible impact on the growth and yield of *Brassica juncea*, in which Pb showed the most adverse effect on growth, followed by Cu and Zn. The plant growth and yield of *B. juncea* decrease with elevated concentrations of heavy metals. Pb decreases the yield by 51% and Cu by 29%, whereas Zn does not show any effect at a concentration of 100 ppm. The yield of *B. juncea* decreases by 38% at a very high concentration of Zn (500 ppm).

1.3 CONCLUSION

This chapter gives a brief introduction for this book. Composting of solid wastes is a good technique to manage solid wastes in a hygienic way and to recycle nutrients. This technique is inexpensive in relation to other waste management techniques.

We can obtain compost from various solid wastes through composting process. However, the presence of the bioavailability of heavy metals in the final compost is a critical issue of composting. Hence, the main focus of this book is to give an overview on the reduction of the bioavailability and leachability of heavy metals during the composting of different waste materials. Composts with elevated concentrations of heavy metals may affect the quality of soil by changing the physicochemical and biological properties of soil. Accumulation of metals by plants from the soil not only decreases the crop productivity but also hinders the physiological metabolism. Transfer of heavy metals in human tissues through the food chain may create problems for human being. Consequently, compost should be free from pathogens and heavy metals before being used in agriculture.

REFERENCES

Ashraf, R., and Ali, T.A. 2007. Effect of heavy metals on soil microbial community and mung beans seed germination. *Pakistan Journals of Botany* 39(2): 629–636.

Bhattacharyya, P., Chakrabarti, K., Chakraborty, A., Tripathy S., and Powell, M.A. 2008. Fractionation and bioavailability of Pb in municipal solid waste compost and Pb uptake by rice straw and grain under submerged condition in amended soil. *Geosciences Journal* 12(1): 41–45.

Bragato, G., Leita, L., Figliolia, A., and Nobili, M. 1998. Effects of sewage sludge pretreatment on microbial biomass and bioavailability of heavy metals. *Soil and Tillage Research* 46: 129–134.

Chen, G.Q., Chen, Y., Zeng, G.M., Zhang, J.C., Chen, Y.N., Wang, L., and Zhang, W.J. 2010. Speciation of cadmium and changes in bacterial communities in red soil following application of cadmium-polluted compost. *Environmental Engineering Science* 27(12): 1019–1026.

Chiang, K.Y., Huang, H.J., and Chang, C.N. 2007. Enhancement of heavy metal stabilization by different amendments during sewage sludge composting process. *Journal of Environmental Management* 17(4): 249–256.

Cu, N.X. 2015. Effect of heavy metals on plant growth and ability to use fertilizing substances to reduce heavy metal accumulation by *Brassica Juncea* L. Czern. *Global Journal of Science Frontier Research: D Agriculture and Veterinary* 15(3): 34–40.

Deka, H., Deka, S., Baruah, C.K., Das, Hoque, J.S., Sarma, H., and Sarma, N.S. 2011. Vermicomposting potentiality of *Perionyx excavatus* for recycling of waste biomass of *Java citronella*—An aromatic oil yielding plant. *Bioresource Technology* 102: 11212–11217.

Fang, M., and Wong, J.W.C. 1999. Effects of lime amendment on availability of heavy metals and maturation in sewage sludge composting. *Environmental Pollution* 106: 83–89.

Fuentes, A., Llorens, M., Saez, J., Aguilar M.I., Marın, A.B.P., Ortuno, J.F., and Meseguer V.F. 2006. Ecotoxicity, phytotoxicity and extractability of heavy metals from different stabilised sewage sludges. *Environmental Pollution* 143: 355–360.

Gabhane, J., William, S.P.M.P., Bidyadhar, R., Bhilawe, P., Anand, D., Vaidya, A.N., and Wate, S.R. 2012. Additives aided composting of green waste: Effects on organic matter degradation, compost maturity, and quality of the finished compost. *Bioresource Technology* 114: 382–388.

Garnier, J., Quantin, C., Martins, E.S., and Becquer, T. 2006. Solid speciation and availability of chromium in ultramafic soils from Niquelandia, Brazil. *Journal of Geochemical Exploration* 88: 206–209.

Garrido, S., Campo, G.M.D., Esteller, M.V., Vaca, R., and Lugo, J. 2002. Heavy metals in soil treated with sewage sludge composting, their effect on yield and uptake of broad bean seeds (*Vicia faba L.*). *Water, Air, and Soil Pollution* 166: 303–319.

Guala, S.D., Vega, F.A., and Covelo, E.F. 2010. The dynamics of heavy metals in plant-soil interactions. *Ecological Modelling* 221: 1148–1152.

He, M., Tian, G., and Liang, X. 2009. Phytotoxicity and speciation of copper, zinc and lead during the aerobic composting of sewage sludge. *Journal of Hazardous Materials* 163: 671–677.

Hinojosa, M.B., Carreira, J.A., Ruız, R.G., and Dick, R.P. 2004. Soil moisture pre-treatment effects on enzyme activities as indicators of heavy metal contaminated and reclaimed soils. *Soil Biology and Biochemistry* 36: 1559–1568.

Hsu, J.H., and Lo, S.L. 2001. Effects of composting on characterization and leaching of copper, manganese, and zinc from swine manure. *Environmental Pollution* 114: 119–127.

Iwegbue, C.M.A., Emuh, F.N., Isirimah, N.O., and Egun, A.C. 2007. Fractionation, characterization and speciation of heavy metals in composts and compost-amended soils. *African Journal of Biotechnology* 6(2): 67–78.

Jordao, C.P., Nascentes, C.C., Cecon, P.R., Fontes, R.L.F., and Pereira, J.L. 2006. Heavy metal availability in soil amended with composted urban solid wastes. *Environmental Monitoring and Assessment* 112: 309–326.

Kang, J., Zhang, Z., and Wang, J.J. 2011. Influence of humic substances on bioavailability of Cu and Zn during sewage sludge composting. *Bioresource Technology* 102: 8022–8026.

Karaca, A., Cetin, S.C., Turgay, O.C., and Kizilkaya, R. 2010. Effects of heavy metals on soil enzyme activities, in I. Sherameti and A. Varma (eds.) *Soil Heavy Metals*, Soil Biology, Vol. 19, pp. 237–262, Springer, Berlin, Heidelberg.

Khan, S., Cao, Q., Zheng, Y.M., Huang, Y.Z., and Zhu, Y.G. 2008. Health risks of heavy metals in contaminated soils and food crops irrigated with wastewater in Beijing, China. *Environmental Pollution* 152: 686–692.

Li, X.D., Poon, C.S., Sun, H., Lo, I.M.C., and Kirk, D.W. 2001. Heavy metal speciation and leaching behaviors in cement based solidified/stabilized waste materials. *Journal of Hazardous Materials* 82: 215–230.

Li, L., Xu, Z., Wu, J., and Tian, G. 2010. Bioaccumulation of heavy metals in the earthworm Eisenia fetida in relation to bioavailable metal concentrations in pig manure. *Bioresource Technology* 101: 3430–3436.

Mora, A.P., Calvo, J.J.O., Cabrera, F., and Madejon, E. 2005. Changes in enzyme activities and microbial biomass after "in situ" remediation of a heavy metal-contaminated soil. *Applied Soil Ecology* 28: 125–137.

Nair, A., Juwarkar, A.A., and Devotta, S. 2008. Study of speciation of metals in an industrial sludge and evaluation of metal chelators for their removal. *Journal of Hazardous Materials* 152: 545–553.

Neklyudov, A.D., Fedotov, G.N., Ivankin, A.N. 2008. Intensification of composting processes by aerobic microorganisms: A review. *Applied Biochemistry and Microbiology* 44(1): 6–18.

Rascio, N., and Izzo, F.N. 2011. Heavy metal hyperaccumulating plants: How and why do they do it? And what makes them so interesting? *Plant Science* 180: 169–181.

Sharma, R.K., Agrawal, M., and Marshall, F. 2007. Heavy metal contamination of soil and vegetables in suburban areas of Varanasi, India. *Ecotoxicology and Environmental Safety* 66: 258–266.

Shun-hong, H., Bing, P., Zhi-hui, Y., Li-yuan, C., and Li-cheng, Z. 2009. Chromium accumulation, microorganism population and enzyme activities in soils around chromium-containing slag heap of steel alloy factory. *Transactions of Nonferrous Metals Society of China* 19: 241–248.

Singh, J., and Kalamdhad, A.S. 2011. Effects of heavy metals on soil, plants, human health and aquatic life. *International Journal of Research in Chemistry and Environment* 1(2): 15–21.
Singh, J., and Kalamdhad, A.S. 2012. Concentration and speciation of heavy metals during water hyacinth composting. *Bioresource Technology* 124: 169–179.
Singh, J., and Kalamdhad, A.S. 2013a. Assessment of bioavailability and leachability of heavy metals during rotary drum composting of green waste (water hyacinth). *Ecological Engineering* 52: 59–69.
Singh, J., and Kalamdhad, A.S. 2013b. Effects of lime on bioavailability and leachability of heavy metals during agitated pile composting of water hyacinth. *Bioresource Technology* 138: 148–155.
Singh, J., and Kalamdhad, A.S. 2014. Influences of natural zeolite on speciation of heavy metals during rotary drum composting of green waste. *Chemical Speciation and Bioavailability* 26(2): 1–11.
Smith, S.R. 2009. A critical review of the bioavailability and impacts of heavy metals in municipal solid waste composts compared to sewage sludge. *Environment International* 35: 142–156.
Speira, T.W., Kettlesb, H.A., Percivalc, H.J., and Parshotam, A. 1999. Is soil acidification the cause of biochemical responses when soils are amended with heavy metal salts? *Soil Biology and Biochemistry* 31: 1953–1961.
Sprynskyy, M., Kosobucki, P., Kowalkowski, T., and Buszewsk, B. 2007. Influence of clinoptilolite rock on chemical speciation of selected heavy metals in sewage sludge. *Journal of Hazardous Materials* 149: 310–316.
Su, Ch., Jiang, L.Q., and Zhang, W.J. 2014. A review on heavy metal contamination in the soil worldwide: Situation, impact and remediation techniques. *Environmental Skeptics and Critics* 3(2): 24–38.
Swati, A., and Hait, S. 2017. Fate and bioavailability of heavy metals during vermicomposting of various organic wastes—a review. *Process Safety and Environmental Protection* 109: 30–45. DOI: 10.1016/j.psep.2017.03.031.
Tchounwou, P.B., Yedjou, C.G., Patlolla, A.K., and Sutton, D.J. 2012. Heavy metals toxicity and the environment. *EXS* 101: 133–164. DOI: 10.1007/978-3-7643-8340-4_6.
Tessier, A., Campbell, P.G.C., and Bisson, M. 1979. Sequential extraction procedures for the speciation of particulate trace metals. *Analytical Chemistry* 51: 844–851.
Villasenor, J., Rodriguez, L., and Fernandez, F.J. 2011. Composting domestic sewage sludge with natural zeolites in a rotary drum reactor. *Bioresource Technology* 102(2): 1447–1454.
Wang, Y.P., Shi, J.Y., Wang, H., Li, Q., Chen, X.C., and Chen, Y.X. 2007. The influence of soil heavy metals pollution on soil microbial biomass, enzyme activity, and community composition near a copper smelters. *Ecotoxicology and Environmental Safety* 67: 75–81.
Wang, X., Chen, L., Xia, S., and Zhao, J. 2008. Changes of Cu, Zn, and Ni chemical speciation in sewage sludge co-composted with sodium sulfide and lime. *Journal of Environmental Sciences* 20: 156–160.
Wang, L., Zheng, Z., Zhang, Y., Chao, J., Gao, Y., Luo, X., and Zhang, J. 2013. Biostabilization enhancement of heavy metals during the vermiremediation of sewage sludge with passivant. *Journal of Hazardous Materials* 244–245: 1–9.
Wong, J.W.C., and Fang, M. 2000. Effects of lime addition on sewage sludge composting process. *Water Research* 34(15): 3691–3698.
Wong, J.W.C., and Selvam, A. 2006. Speciation of heavy metals during co-composting of sewage sludge with lime. *Chemosphere* 63: 980–986.
Yao, H., Xu, J., and Huang, C. 2003. Substrate utilization pattern, biomass and activity of microbial communities in a sequence of heavy metal polluted paddy soils. *Geoderma* 115: 139–148.

Yuan, X., Huang, H., Zeng, G., Li, H., Wang, J., Zhou, C., Zhu, H., Pei, X., Liu, Z., and Liu, Z. 2011. Total concentrations and chemical speciation of heavy metals in liquefaction residues of sewage sludge. *Bioresource Technology* 102, 4104–4110.

Zheng, G.D., Chen, T.B., Gao, D., and Luo, W. 2007. Stabilization of nickel and chromium in sewage sludge during aerobic composting. *Journal of Hazardous Materials* 142: 216–221.

Zojaji, F., Hassani, A.H., and Sayadi, M.H. 2014. Bioaccumulation of chromium by *Zea mays* in wastewater-irrigated soil: An experimental study. *Proceedings of the International Academy of Ecology and Environmental Sciences* 4(2): 62–67.

Zorpas, A.A., Constantinides, T., Vlyssides, A.G., Haralambous, I., and Loizidou, M. 2000. Heavy metal uptake by natural zeolite and metals partitioning in sewage sludge compost. *Bioresource Technology* 72: 113–119.

2 Composting Process

2.1 COMPOSTING PROCESS

The composting process can be defined as the aerobic decomposition/conversion of organic matter by different microorganisms into CO_2, H_2O (water vapor form), NH_3, nutrients (such as nitrogen, phosphorus, and potassium), and stable humic-like substances (Haroun et al., 2009). It is a highly economical method to manage organic wastes generated from numerous sources such as sewage sludge, animal manure, industrial wastes, and agricultural waste (Haroun et al., 2009). Oxygen is mandatory for decomposition of organic matter under the aerobic condition (Figure 2.1). The main products obtained after biological metabolism under this condition are carbon dioxide, water in the form of water vapor, and heat.

2.2 FACTORS AFFECTING THE COMPOSTING PROCESS

One of the most important parameters affecting the composting process is the carbon-to-nitrogen (C/N) ratio. Phosphorus is the second most important parameter followed by sulfur, calcium, and trace metals. These parameters play a vital role in many aspects of cellular metabolism.

2.2.1 Carbon-to-Nitrogen (C/N) Ratio

The C/N ratio of the organic material to be composted is important as it affects the microbial community in the composting mixture, stabilization of the final product, and the availability of nutrients. The composting time and the cumulative CO_2 production have been shown to be linearly dependent on the initial C/N ratio (Chang and Hsu, 2008). The initial C/N ratio is one of the most important factors that strongly prompt the quality of compost (Guo et al., 2012a). When the C/N ratio is higher, the composting process takes more time to stabilize the organic material, and the microbial community will also tend to an increase in abundances of fungal biomass (Eiland et al., 2001). Nitrogen is limiting in materials with a C/N ratio that is too high (>100) (Leconte et al., 2009).

The microorganisms responsible for the stabilization of organic matter have a C/N ratio of 30:1. Therefore, the C/N ratio of initial composting materials ranging from 25 to 30 is considered most satisfactory for the composting process (Kumar et al, 2010). Some researchers found that the optimum value varies from 26 to 31 depending on the environmental conditions (Kalamdhad, 2010). However, some researchers have successfully conducted the composting process at lower initial C/N ratios as follows: 19.6 for the green waste and food waste composting (Kumar et al., 2010), 20 for the composting of chicken manure with sawdust (Ogunwande et al., 2008), 20 for the composting of swine manure with

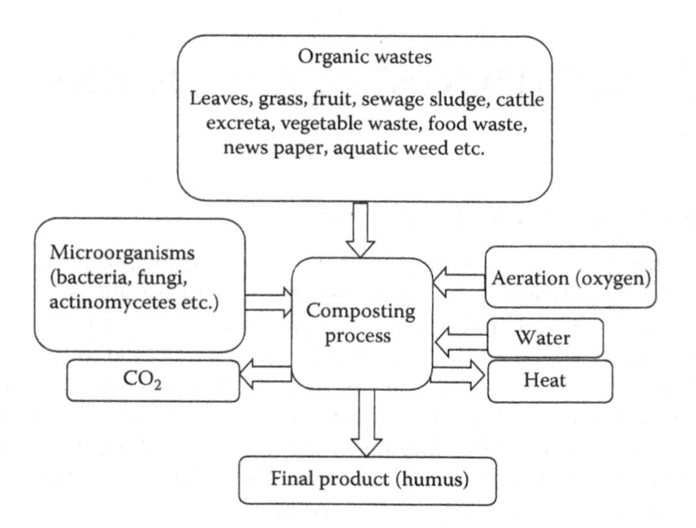

FIGURE 2.1 Outline of the composting process.

rice straw (Zhu, 2007), and 15 for the composting of pig manure with sawdust (Huang et al., 2004).

2.2.2 Particle Size

For efficient aeration of composting materials, a small particle size is required to provide supportive conditions to bacteria, fungi, and actinomycetes. These microorganisms are highly efficient decomposed small size of organic materials. Therefore, a larger size of municipal solid waste (MSW) and agricultural wastes should be shredded into small pieces (25 and 75 mm). Sewage sludge, cattle manure, etc. typically contain fine solid particles. These materials are appropriate for degradation by microbial community (Kalamdahd, 2010).

2.2.3 Moisture Content

The moisture content in the biomass to be composted is an important parameter. The moisture content of the composting mixture generally occupies the free air space between the particles. Higher moisture content may create anaerobic conditions in the compost mass. However, the composting mass should have a certain minimum moisture content in it for the organisms to survive. Sufficient moisture content is required for composting of straw and other materials holding strong fibrous, after absorption of water fiber became soft and fills the large pore spaces among the various particles (Kalamdhad, 2010). The microbial activity as well as the physical structure can be affected by moisture content, and thus plays a vital role in the biodegradation of organic materials (An et al., 2012). It has been reported that the microbial activity is inhibited when the moisture content is <25%, and the aeration can be restricted when the moisture content is >70% (Rodriguez et al., 1995). Most materials are best composted in a moisture content ranging from 50% to 70%.

2.2.4 Porosity

Porosity, which is largely affected by moisture content, is an important parameter because it directly affects O_2 availability within the compost piles. Air flows through a compost pile by the network formed by air-filled pores, so it is important to distinguish air-filled porosity (AFP) from total porosity (Ruggieri et al., 2009). Porosity (ε) is defined as the ratio of the total void volume of the sample (V_v), comprised of air and water filled voidage, to the total volume of the sample (V_s) (Han et al., 2014), as shown in Eq. (2.1). AFP is defined as the ratio of the air volume (V_g) to the total volume of the sample (V_s) (Haug, 1993), as shown in Eq. (2.2).

$$\varepsilon = \frac{V_v}{V_s}, \tag{2.1}$$

$$\text{AFP} = \frac{V_g}{V_s}. \tag{2.2}$$

The lowest AFP values in freshly prepared organic material that will ensure the aerobic microbial activity has been reported to be 30% (Jeris and Regan, 1973; Haug, 1993), whereas the AFP values over 60%–70% seem to be excessive to achieve thermophilic temperatures in wastes with low-biodegradable organic matter content.

2.2.5 Oxygen Requirement

It is well known that oxygen is required for the microorganisms to maintain aerobic activities in the composting mixture. The amount of oxygen gets exhausted and it has to be uninterruptedly supplied to the microorganisms to decompose organic materials. A continuous supply of oxygen can be attained either by turning windrows or by supplying air through compressor. This is essential to transport inner mass to the outer surface and to transfer the outer mass to the inner portion during the turning of the composting mixture. The amount of air supply is usually sustained at 1–2 m^3/day/kg of volatile solids for air supply through compressor. Approximately 484–674 kcal energy is being released by aerobic decomposition of 1 g mole of glucose molecule under the controlled conditions, whereas only 26 kcal is produced when a glucose molecule is decomposed in an oxygen-free environment (Kalamdhad, 2010). The aeration rate is one of the most important factors affecting the composting process (Diaz et al., 2002). An inadequate air supply will promote anaerobic conditions due to the deficiency of oxygen, although extreme aeration will increase costs and slow down the composting process through loss of heat, water, and ammonia (Guo et al., 2012b).

2.2.6 Temperature

Temperature is essential to the removal of non-spore-forming pathogens, for example, *Salmonella* and *Escherichia coli*. It is recommended that in the composting process, the temperature must exceed 55°C for a period of at least 2 weeks (Han et al., 2014). In the thermophilic phase of composting, the temperature is increased due to the release of exothermic energy in the process of oxidation of organic matter. The level of temperature rise depends on the rate of metabolic activity, the extent of oxidation, and the rate of heat transfer from the composting material. Based on the results obtained through theoretical analysis, there is a positive correlation between temperature rise and CO_2 production (Tang et al., 2011). If sufficient nutrients and moisture are present in composting mixture, temperatures will rise >70°C in the composting process due to degradation of organic materials aerobically. Thermophilic bacteria are developed when the temperature goes beyond 40°C, and then the composting process is controlled by thermophilic bacteria: the most dominant species of thermophilic bacteria is *Bacillus*. The highest diversity of bacilli species can be measured at temperatures in the range of 50°C–55°C; however, the diversity of these bacteria decreases intensely at a temperature of 60°C or >60°C. Furthermore, in unfavorable conditions, these bacteria survive through formation of thick-walled endospores, which are extremely impervious to environmental conditions such as heat, cold, dryness, and food scarcity. Bacterial endospores are ubiquitous in nature and develop in the active form when they are under favorable environmental conditions (Kalamdhad, 2010). The activity of cellulose enzymes drastically decreases at 70°C or >70°C. The optimum temperature, ranging from 30°C to 50°C, is highly recommended for nitrification, whereas the temperature over this range is not supposed to be good for nitrification (Kalamdhad, 2010).

The thermophilic temperature is used for the destruction of pathogens as well as parasites through the death of a cell by means of thermal inactivation of its enzymes.

According to Haug (1993), pathogens are killed by antagonism with other microbes and by antibiotics produced by microbes, which inhibit growth of other microbes. Therefore, the US Environmental Protection Agency (EPA) suggested different temperature levels for killing pathogens in the composting mass: 53°C for 5 days, 55°C for 2 days, and 70°C for 30 minutes.

2.2.7 pH Value

pH is also one of the most important parameters that significantly affect the composting process. The pH values are suitable for the development of bacteria and fungi are in the range of 6.0–7.5 and 5.5–8.0, respectively (Amir et al., 2005). Zhu (2007) stated that the pH value tended to be stable and usually seemed to be constant in all the composts. It increased significantly in the first week of composting, then slowly decreased, and increased faintly in the maturation phase. The pH values were found to be 7.30 and 7.36 in the initial composting mixtures of bin 1 and bin 2, respectively, whereas the pH values were found to be 8.01 and 8.03 in the final compost. Singh and Kalamdhad (2016) reported that with the addition of lime, the pH value of the initial composting mass was higher compared to that without lime. However, it decreased progressively to approximately neutral. The pH value of the initial composting mixture (lime added) was found to be 10.4, whereas the pH value of the final compost was found to be 7.7.

2.3 TYPES OF COMPOSTING PROCESS

Figure 2.2 shows the different types of compositing processes.

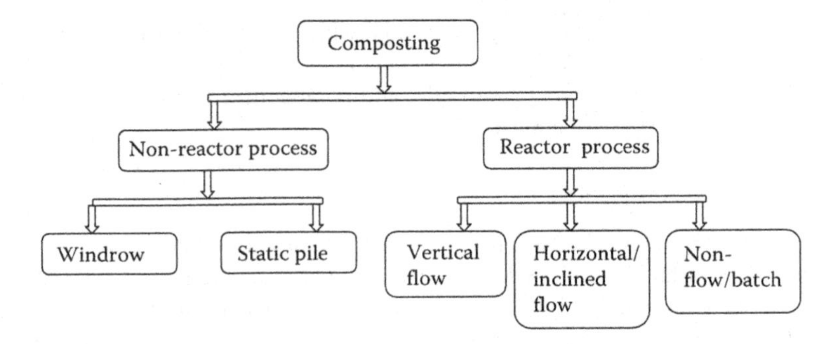

FIGURE 2.2 Different types of compositing processes.

2.3.1 Agitated Pile/Windrow

The agitated pile bed system is the most popular example of a nonreactor composting. Feedstocks of various waste mixtures are located in different rows, and these rows are turned occasionally manually or by using mechanical equipment. Height, width, and shape of the windrows can be determined based on the nature of the feed material and the turning equipment to be used. Oxygen can be supplied mainly through natural ventilation and by gas exchange when the pile is turning. The periodic agitation of windrow is not only beneficial for restructuring the windrow but it also provides proper aeration to microbes. Consequently, significant mixing can be predictable along the height and width of the row. The height of the composting pile is decreased during the progression of the composting process due to mass loss through the degradation of organic materials.

Agitated pile composting mixtures can be obtained through trapezoidal piles (height of 550mm, length of 2100mm, base width of 350mm, and top width of 100mm) with a length-to-base width (L/W) ratio of 6. Figure 2.3 shows a schematic diagram and a pictorial view of the agitated pile. Agitated piles contain

FIGURE 2.3 Pictorial view of agitated pile composting.

approximately ≥150 kg composting materials with different waste combinations (Singh and Kalamdhad, 2012). After turning the piles, grab samples can be collected from different locations of the agitated pile, and homogenized samples are prepared after proper mixing of waste materials.

2.3.2 Aerated Static Pile

The aerated static pile process is a nonreactor stationary solid bed system, as shown in Figure 2.4. Similar to agitated piles, the waste materials are mixed properly with bulking agents (e.g., wood chips, sawdust, etc.) and is formed into a large heap. The bulking agent provides the structural stability to the material and maintains air voids without the need for periodic agitation. An air supply system is used to allow either forced or induced draft aeration.

No agitation or turning of the compost biomass occurs during the composting process, and the static piles can be formed on a batch basis. Since there is no bed agitation, no mixing occurs once the pile is formed.

2.3.3 Vertical Flow Reactor

Vertical flow reactors may be well defined according to the bed conditions in the reactor. The examples of vertical reactor are agitated solid bed reactor and packed bed reactor (silo reactor). Some systems allow for agitation of solids during their transit down the reactor and are termed as moving agitated bed reactors. These reactors are frequently fed on either a continuous or intermittent basis. In other reactors, the composting mixture occupies the entire bed volume and is not agitated during any single pass in the bed. These reactors are called moving packed bed reactors and can be fed on a continuous, an irregular, or a batch basis. The moving packed bed systems often allow for periodic transfer of solids from the bottom to the top of the reactor.

Agitation of solids occurs as a result of this transfer, but on any single pass, the bed solids remain static until they are again withdrawn from the bottom for transfer.

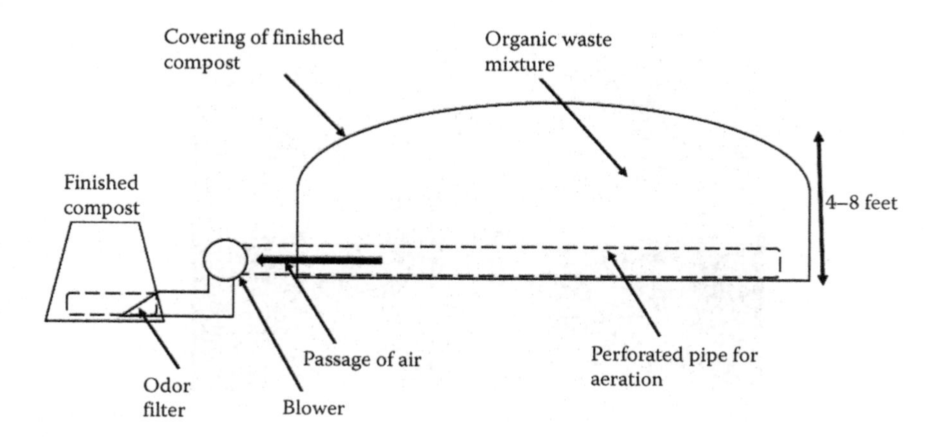

FIGURE 2.4 Schematic view of aerated static pile composting.

2.3.4 Horizontal Flow Reactor (Rotary Drum Reactor)

The best example for horizontal flow reactors is the rotary drum reactor. Horizontal flow reactors are divided into those that employ a rotating or rotary drum, those that use a bin structure of varying geometry and method of agitation (agitated solids bed reactors), and those that use a bin-type structure but with a static solids bed (static solids bed reactors). Such reactors have been used to a wide variety of composting substrates, such as MSW, agricultural wastes, and sewage sludges (Kalamdhad et al., 2009). Rotary drums can be distinguished based on the solids flow pattern within the reactor. In the dispersed flow system, material inlet and outlet are located on the opposite ends of the drum. Plug flow conditions exist within the vessel. This type of reactor is probably the most commonly used drum system and has been widely applied to rotary drum systems.

The rotary drum composter is a highly effective and encouraging technique in decentralized composting of waste materials. This reactor delivers agitation, aeration, and mixing of the compost, resulting in an end product homogeneously with odor or leachate problems (Kalamdhad et al., 2009). Figure 2.5 shows the pictorial view of a rotary drum composter of 550L capacity, which can be operated in a batch-type manner. The main unit of the composter, i.e., the rotary drum of 1.22m in length and 0.76m in diameter, is made by a 4 mm-thick metal sheet. The inner surface of the composter is coated with anticorrosive materials. This reactor may be riding on four rubber rollers that are attached to a metal stand, and it can also be turned manually with the handle attached with it. For proper mixing of composting mass, 40 × 40mm angles are generally welded longitudinally inside the composter. Two adjacent holes of 10cm each are made on the bottom of the composter to drain out the excess water.

The composting time suggestively reduced to 2–3 weeks by providing optimum conditions for microorganisms. These conditions are optimum temperature and moisture with the sufficient amount of food and oxygen inside the rotary drum composter. These conditions will allow the aerobic microbes to grow and decompose the waste rapidly.

FIGURE 2.5 Pictorial view of a rotary drum composter.

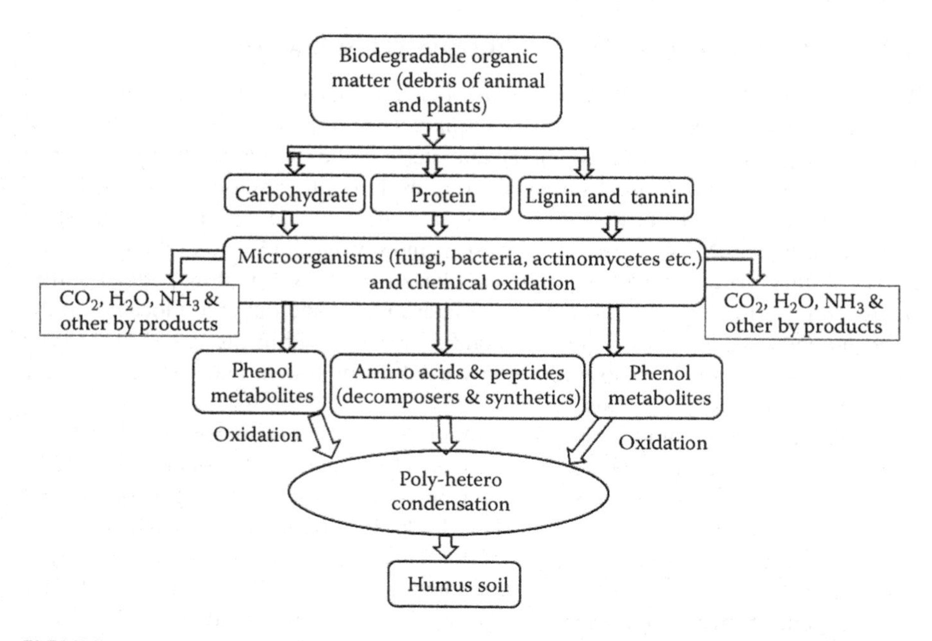

FIGURE 2.6 Formation of natural humus soil from the debris of plants and animals.

Rotary drum reactors have been applied to composting a variety of waste materials (cattle manure, swine manure, municipal biosolids, brewery sludge, chicken manure, olive mill waste, sewage sludge, tannery sludge, and food residuals) (Kalamdhad et al., 2009; Fernandez et al., 2010; Rodriguez et al., 2012). A mechanism of humus formation naturally from the debris of plants and animals (Figure 2.6), which contains humic substances such as organic macromolecule, which hold high structural complexity in their structure (Chaturvedi et al., 2006; Zhu et al., 2011).

A number of composting technologies that use batch operated compost "boxes" are available. The composting mixture is loaded at the start of the cycle and typically remains in the box reactor. Aeration is usually controlled and may alternate between positive and negative modes. Curing is commonly done in the windrow system for several months.

2.4 PHASES OF THE COMPOSTING PROCESS

When the heat produced by the metabolic activities of microorganisms is prevented by some kind of insulation from being dissipated to the environment, the temperature of the habitat of microorganisms increases. The organic matter is collected in bulky heaps or kept in containers when the organic waste is composted either in large agitated piles (windrows) or in composters (Kutzner, 2001).

These thermophilic phases considered for the activities of successive microbial populations perform the decomposition of more refractory organic matter. As shown in Figure 2.7, the time–temperature trend of the composting process may be classified into four phases (Kutzner, 2001):

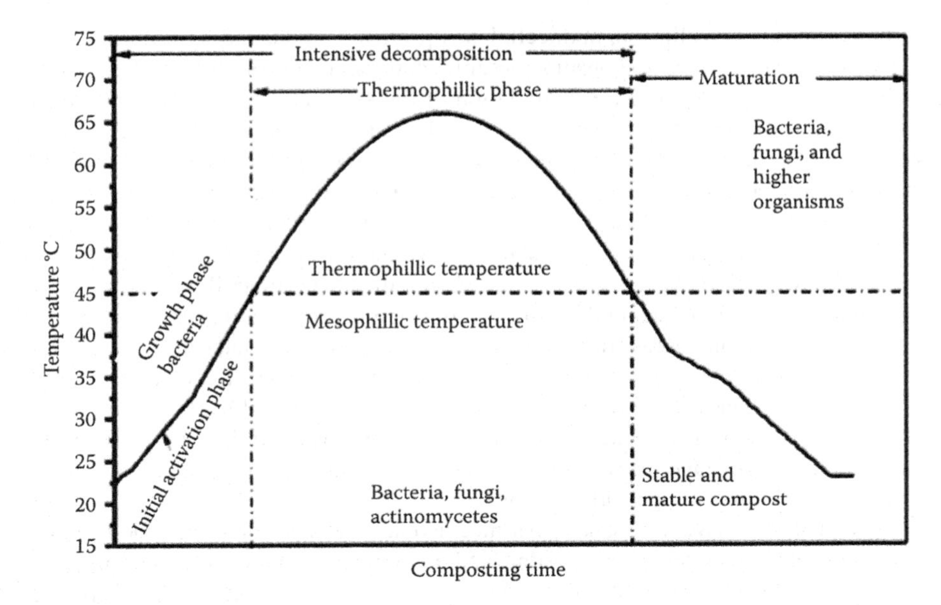

FIGURE 2.7 Various phases of the composting process.

1. During the first phase of composting, a diverse population of mesophilic bacteria and fungi proliferates. These microorganisms are involved in degrading the readily available nutrients and increasing the temperature to about 45°C. At this temperature level, the microbial activities stop growing and only heat-resistant spores survive.
2. The second phase is characterized by the development of a thermophilic microbial population comprising some bacterial species, actinomycetes, and fungi. These microorganisms grow well at an optimum temperature of 50°C–65°C; however, their activities terminate at >70°C.
3. The third phase is generally known as a stationary phase. In this phase, no significant changes of temperature occur because microbial heat production and heat dissipation balance each other. The microbial population continues to consist of thermophilic bacteria, actinomycetes, and fungi.
4. The fourth phase is characterized by a reduction of a steady-state temperature. It is best labeled as the maturation phase of the composting process. The mesophilic microorganisms may survive at even higher temperatures during the thermophilic phase.

2.5 MICROORGANISMS INVOLVED DURING COMPOSTING

During composting, microorganisms transform the organic matter into CO_2, biomass, energy in the form of heat, and a humus-like final product. Mostly plant-derived materials, such as organic substrates, bulking agents, and amendments, are used in the composting process. Organic substrates are mainly composed of carbohydrates (e.g., cellulose, hemicellulose), proteins, lipids, and lignin. The capability of microorganisms to consume organic matter depends on their proficiency to yield the enzymes required for degradation of the substrate. The more widespread and inclusive the enzyme system required to degrade the more complex the organic substrate. Complex organic compounds break down into smaller molecules through the synergistic action of microorganisms. These microorganisms can utilize smaller molecules of organic substances by the microbial cells (Tuomela et al., 2000).

Various microorganisms required for the formation of compost such as bacteria, fungi, and actinomycetes grow under mesophilic or thermophilic environments. Microbial communities depend on the composition of the organic substrate, which is involved in the biodegradation in the composting process. Bacterial communities involved in the aerobic degradation of organic waste are shown in Table 2.1. The highest treating efficiency is achieved by combining the activities of bacteria with those of protease and cellulose. Exact examples of such bacteria and their involvement in the complex, multistage process of degradation (and subsequent mineralization) of composted mixtures are provided below.

Several studies have represented that inoculation is satisfactory for the biodegradation of organic matter during composting, and the composting process can be enhanced (Shukla et al., 2009; Tang et al., 2011; Ridha et al., 2012). Sutripta et al. (2010) reported that inoculation by thermophilic bacteria could be effective in the thermophilic phase of the composting process. Nakasaki et al. (2013) reported that the inoculation of *Pichia kudriavzevii* RB1 improved the degradation of organic acids present in the raw compost material (food waste) and accelerated the composting process. Hachicha et al. (2012) reported that the inoculation of *Trametes versicolor* improved the degradation of lignin resulting in the enhanced humification process. Cao et al. (2013) reported that the inoculation of *Aspergillus fumigatus* F12 improved the degradation of cellulose prior to other biopolymers. Xue et al. (2010) stated that the inoculation of nitrifying bacteria augmented NH_3 removal during composting.

2.5.1 Bacteria

Bacteria are generally classified as unicellular microorganisms with a size of 0.5–3.0 μm. They have a very high surface-to-volume ratio because of their small size, resulting rapid allocation of soluble substrates into the bacterial cell. Bacteria are generally more dominant than larger microorganisms such as fungi. A few bacteria, for example, *Bacillus* spp., have the capability to produce thick-walled endospores. They are highly stable to heat, harmful radiation, and chemical disinfection (Haug, 1993). A wide range of bacteria have been identified from different compost environments, including the species of *Pseudomonas*, *Klebsiella*, *Bacillus badius* AK, and *Bacillus* sp. HPC40 (Nakasaki et al., 1985; Strom, 1985;; Liu et al., 2011; Vishan et al., 2017a). The three main bacteria of the thermophilic phase are *B. subtilis*, *B. licheniformis*, and *B. circulans*. During the thermophilic phase of composting, approximately 87% of the randomly selected colonies belong to the genus *Bacillus* (Strom, 1985). Numerous thermophilic species of *Thermus* had been isolated from the thermophilic phase of composting at temperatures ranging from 65°C to 82°C (Tuomela et al., 2000).

TABLE 2.1
Key Microorganisms Involved in Aerobic Composting of Different Waste Materials

Waste Types	Microorganisms	References
Microcrystalline cellulose	*Fomitopsis* *Palustris*	Yoon et al. (2007)
Wood	*Xylaria polymorpha*	Xing-Na et al. (2005)
Microcrystalline cellulose	*Thermobifida fusca*, *Geobacillus* sp., *Bacillus* sp., and *Geobacillus* sp.	Rastogi et al. (2009) Rastogi et al. (2010) Harun et al. (2012)
Agro-waste cocktail	*Bacillus licheniformis* 2D55	Kazeem et al. (2017)
Agricultural waste	*Cladosporium bruhnei*, *Hanseniaspora uvarum*, *Scytalidium thermophilum*, *Tilletiopsis penniseti*, and *Coprinopsis*	Yu et al. (2015)
Food waste	*Pichia kudriavzevii* RB1	Nakasaki et al. (2013)
Lignin	*Trametes versicolor*	Hachicha et al. (2012)
Cellulosic waste	*Aspergillus fumigatus* F12	Cao et al. (2013)
Empty fruit bunches and palm oil mill effluent	*Trichoderma virens*	Amira et al. (2011)
Water hyacinth	*Bacillus badius* AK	Vishan et al. (2017a, 2017b)
Wheat straw waste	*Agaricus subrufescens*	Farnet et al. (2013)
Human waste	*Arthrobacter*, *Streptomyces*, *Actinomadura*, *Thermobifida fusca*, *Saccharomonospora*, *Streptosporangium*, *Bacillus*, *Geobacillus*, *Ureibacillus*, and *Planifilum*	Piceno et al. (2017)
Cow manure compost	*Bacillus* sp. HPC40 (AY803983) *Bacillus thermocloacae* (Z26939) *Pseudomonas* sp. *129(43zx)*	Liu et al. (2011)
Pressmud compost	*Lichtheimia ramose*, *Myceliophthora fergusii*, *Myceliophthora thermophile*, *Rhizomucor pusillus*, *Rhizopus microspores*, *Mycothermus thermophilum*, *Thermomucor indicae-seudaticae*, and *Thermomyces lanuginosus*	Oliveira et al. (2016)

Thermophiles can be characterized into moderate thermophiles (optimum growth temperature at 50°C–60°C). The optimum temperature for the growth of extreme thermophiles is in the range of 60°C–80°C, whereas the optimum temperature for the growth of hyperthermophiles is in the range of 80°C–110°C (Gupta et al., 2014). Thermophilic microorganisms growing at an optimum temperature of ≥50°C are considered as extremophiles due to the generation of thermostable enzymes (such as amylases, cellulases, chitinases, pectinases, xylanases, proteases, lipases, and DNA polymerases). These enzymes show restricted characteristics that can be appropriate for composting processes at higher temperatures (Singh et al., 2011).

Actinomycetes are a type of bacteria that are formed by multicellular filaments. Therefore, these bacteria have some similarities with fungi. Generally, these can be found in the thermophilic phase, cooling phase, and maturation phase of composting, and can occasionally become so frequent that they can be seen on the surface of the compost. Corss (1968) reported that thermophilic actinomycetes had been isolated from a wide range of natural substrates, such as desert sand and compost. On the other hand, some researchers (Waksman et al., 1939; Strom, 1985) reported that the thermophilic actinomycetes were identified from the compost composed of *Nocardia*, *Streptomyces*, *Thermoactinomyces*, and *Micromonospora*. Actinomycetes are capable of degrading some cellulose and solubilizing lignin compared to fungi, and they can survive at higher temperatures and pH. Consequently, actinomycetes are the important agents for the degradation of lignocellulosic biomass during peak heating. Although their capacity to degrade cellulose and lignin is low compared to fungi (Crawford, 1983; Godden et al., 1992), actinomycetes can survive through spore formation under adverse conditions (Cross, 1968).

2.5.2 Fungi

The main factors responsible for the growth of fungi are carbon, nitrogen, pH, and temperature. A reasonably high content of nitrogen is required for the optimum growth of fungi; however, some fungi, which grow on rotten wood, breed at low nitrogen levels. Furthermore, a low level of nutrient (mainly nitrogen) is frequently required to degrade lignin (Dix and Webster, 1995; Tuomela et al., 2000). For the degradation of cellulose, low nutrient nitrogen is a rate-limiting factor (Dix and Webster, 1995). According to Dix and Webster (1995), an acidic environment is favorable for most of the fungi, and these fungi can survive at a wide range of pH values (with exception, *Basidiomycota*o not grow well if pH is >7.5), whereas *Coprinus* species can grow only in the alkaline environment.

Most of the fungi are mesophiles that can grow at temperatures ranging from 5°C to 37°C; however, the optimum temperature for the growth of fungi is in the range of 25°C–30°C (Dix and Webster, 1995). Conversely, during the thermophilic phase of the composting process, a minor group of thermophilic fungi is a significant agent for the degradation of organic matter. The thermophilic fungi can grow in compost heaps in the garden, bird nests, and the storage of numerous agricultural wastes (Dix and Webster, 1995; Tuomela et al., 2000). Ghazifard et al. (2001) reported that the increase in temperature during the thermophilic phase of composting caused an alteration in the biodiversity of microbiota. An increase in temperature resulted in a decrease in the diversity of mesophilic fungi and an increase in the diversity of heat-tolerant fungi. They also reported that the increase in temperature generally led to an increase in the microbial diversity of some species that are resistant to heat, whereas the diversity of mesophilic species and pathogens decreased during the thermophilic phase of pressmud composting. *Phanerochaete chrysosporium* is a well-known white rot fungus which grows at temperatures (favorable) ranging from 36°C to 40°C (Mouchacca, 1997; Tuomela et al., 2000). *Ganoderma colossus* is a different white rot fungus which grows at an optimum temperature of 45°C (Adaskaveg et al., 1995; Tuomela et al., 2000).

2.6 DEGRADATION OF LIGNOCELLULOSIC BIOMASS BY FUNGI-BACTERIA

Lignocellulose is the main constituent of plant biomass containing around half of the plant matter produced by the photosynthesis, demonstrating the most abundant renewable organic resource on earth. It comprises three types of biopolymers: cellulose, hemicellulose, and lignin, which are strongly interlinked and chemically bonded by non-covalent forces and covalent cross-linkages (Pérez et al., 2002; Sánchez, 2009). These biopolymers are generated as by-products in agriculture or forestry, small quantity is used, and the remaining part has been considered as waste biomass. A number of microorganisms present in nature, which are able to degrade cellulose and hemicellulose, can be used as a carbon source of energy. Lignin is one of the most resistant components of plant cell walls which can be degraded to a very small group of filamentous fungi known as white rot fungi, which possess the unique ability to convert lignin effectively to CO_2 and other products (Sánchez, 2009). A thermophilic Ascomycotina, *Thermoascus aurantiacus*, has a high capability to degrade lignin, and this fungus was isolated from the compost (Tuomela et al., 2000). Temperature-resistant soft rot fungi, for example, *Thielavia terrestris*, *Paecilomyces* sp., and *Talaromyces thermophilus*, have less capability to degrade lignin (Dix and Webster, 1995). Figure 2.8 represents the different phases of lignocellulosic biomass bioconversion (Sánchez, 2009).

Organisms such as fungi are mainly accountable for the degradation of lignocellulosic biomass, and basidiomycetes group degrades lignin most quickly compared to the other groups (Rabinovich et al., 2004; Sánchez, 2009). Fungi that have an ability to degrade lignocellulose efficiently are associated with a mycelial growth pattern that allows the fungi to transport the rare nutrients such as nitrogen and iron to a distance where lignocellulosic substrate has poor nutrient and

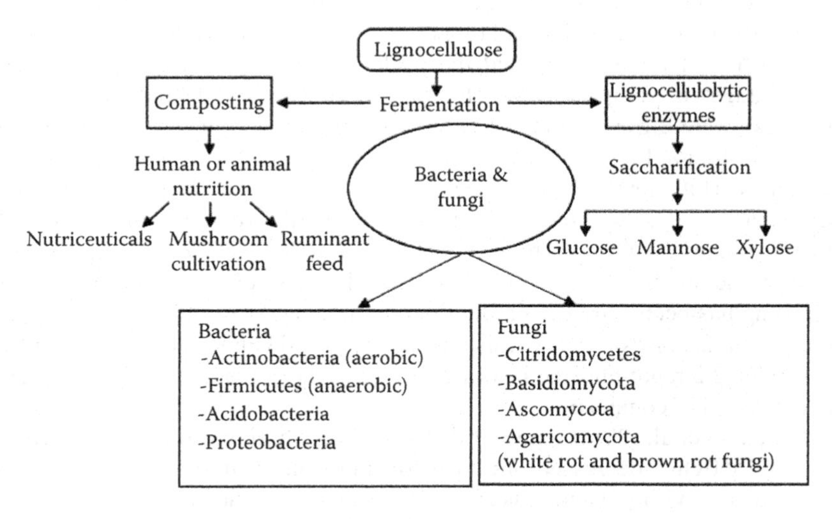

FIGURE 2.8 Different stages of lignocellulosic waste conversion through bacteria and fungi.

only contains the carbon source (Sánchez, 2009). Because of the insolubility of lignin, cellulose, and hemicellulose, the fungal degradation usually occurs exocellularly, either in association with the outer cell covering layer or extracellularly. There are two types of extracellular enzymatic systems found in fungi: (1) the hydrolytic system, which generates hydrolases that are liable for the degradation of polysaccharide and (2) the exclusive oxidative and extracellular ligninolytic system, which is responsible for the degradation of lignin and opens phenyl rings (Sánchez, 2009). Numerous microorganisms, mostly fungi, have been identified and isolated as lignocellulosic degrading agents that are the white rot fungi used for lignocellulolytic enzyme production, for example, *P. chrysosporium* (holobasidiomycetes) and *Trichoderma reesei* (ascomycetes). These fungi are used for the production of hemicellulases and cellulases commercially (Jørgensen et al., 2003; Sánchez, 2009).

2.7 FOURIER-TRANSFORMED INFRARED (FTIR) SPECTRA COMPOST SAMPLES

The composting process encourages the degradation of organic matter in the presence of oxygen, and therefore, chemical transformations are likely to be dissimilar from the organic matter formed after an anaerobic digestion. Conversely, some properties are similar in both cases, such as the degradation of effortlessly degradable molecules and the higher stability of the final product as well as the better chemical complexity in comparison with the starting materials (Provenzano et al., 2014). Hsu and Lo (1999) reported that an increase in aromatic compounds and a decrease in carbohydrates indicate that easily degradable organic matter constituents such as aliphatic and amide components, polysaccharides, and alcohols, are chemically or biologically oxidized, and therefore, the mature compost contains more aromatic structures of higher stability after composting of pig manure for 122 days. Cai et al. (2005) studied the infrared spectra of compost samples collected at different stages of composting. They also reported the peaks observed at 2922 and 2855 cm^{-1} at different stages of composting which assigned aliphatics and lipids. Intensities of these peaks were different and decreased with progression of composting process. The peaks observed at 1540 cm^{-1} represent peptide structures, whereas the peaks observed at 1140–1090 cm^{-1} represent carbohydrates. The intensity of the peak observed at 1384 cm^{-1} was increased showing aromatic structures with phenol (OH) and carboxylates (–COO–). The peaks observed at 1034–1027 cm^{-1} showed etherified aromatic structures. This study concluded that sludge decomposition during composting has been started with the lipid, protein, and carbohydrate components, leading to an increase of highly resistant and stabilized humic substances (Cai et al., 2005). Table 2.2 represents the infrared spectra of the compost samples collected at different stages of composting.

Provenzano et al. (2014) reported that the peaks observed in the polysaccharide region at 1560 cm^{-1} decreased due to biodegradation of amino chain and sugars. On the other hand, the bands associated with fats and lipids at 2920–2930 and at 2851 cm^{-1} decreased due to a reduction in the levels of aliphatic compounds. Ge et al. (2015) reported that the peaks observed at 2925 and 875 cm^{-1} decreased due to the

TABLE 2.2
Infrared Spectra of the Composts of Different Waste Materials

Wave Number (cm^{-1})	Vibration	Functional Group	References
3300–3400	C–H, O–H	H-bonds and OH groups	Provenzano et al. (2014)
2900	C–O–C	Amide region	Tandy et al. (2010)
2925, 2922, and 2855	C–H	Aliphatic and lipid	Amir et al. (2005) and Ge et al. (2015)
2850	C–H stretch	Methylene	Smith (1999)
2520	C=O	Carbonate	Tseng et al. (1996),
1720–1740		Aldehyde, ketone, and carboxylate	Ouatmane et al. (2000), Smith (1999), and Tan (1993)
1640–1650	C=O	Primary amides and aromatics	Haberhauer et al. (1998) and Ouatmane et al. (2000)
1560	C–N	Secondary amides	Provenzano et al. (2014)
1540	C–N	Peptide structure	Amir et al. (2005)
1512		Lignocellulose	Ouatmane et al. (2000)
1430	CO_2 stretch	Carboxylic acid	Smith (1999)
1432	O–H	Phenolic group	Provenzano et al. (2014)
1384	N–O	Nitrate	Smidt et al. (2002)
1320	C–N	Aromatics and amines	Smith (1999) and Ge et al. (2015)
1250–1200		Phospholipid	Tandy et al. (2010)
1260–1240	C–O	Carboxylic acids	Smith (1999)
1160 and 1080	C–O–C	Polysaccharides	Grube et al. (2006)
1030	Si–O–Si	Silica	Huang et al. (2006)
1040	C–O	Polysaccharides	
875	C–O	Carbonates and polysaccharides	Bosch Reig et al. (2002) and Ge et al. (2015)

C–H stretching of aliphatics and the C–O stretching of polysaccharides, respectively, thus representing the degradation of aliphatics and polysaccharides in pig manure composting. The peak observed at 1320 cm^{-1} assigned to the N–H stretching shows aromatics structures. This peak signifying the formation of stable compost and the decline of microbial activities at maturity of composting

2.8 ADVANTAGES OF COMPOSTING

The main advantages of composting of solid wastes are as follows: (1) the production of a marketable product, (2) the reduction of offensive odors caused by rotting waste, (3) a decrease in environmental pollution caused by solid waste, (4) a decrease in the volume and weight of waste materials as a result of drying and degradation of organic matter, and (5) an increase in the nutrient concentration in the final compost of waste due to the weight loss of waste (Oliveira et al., 2016). Composting is used for reducing the municipal solid waste in urban cities through a recycling process of organic waste. Adding compost to soil will increase the nutrient content necessary for plant growth and thus reduces the dependency of chemical fertilizers. Composting is also a low-cost method for the recycling of organic waste. Soil pH may be changed with the addition of compost. The pH values ranging from 6.0 to 7.5 are considered to be optimum in the cultivation of fruits, vegetables, and herbaceous ornamental plants. When the pH of the soil is >7.5, the added compost will decrease the soil fertility, whereas it will improve the soil fertility when the soil has a pH level of <7 under the acidic condition. Composting is used to recycle the organic waste materials from fruits and vegetables, sewage sludge, and other organic wastes in a composting pile, thus preventing nutrient to the landfilling. Compost has the unique ability to improve soil quality and structure. When compost is added to the soil, the nutrient content and moisture-holding capacity increases. An important benefit of the composting process is that its high temperature basically kills all pathogens and weed seeds that may be present in wastes (Kalamdhad et al., 2009).

2.9 DISADVANTAGES OF COMPOSTING

Composting has many advantages and few disadvantages. Disadvantages of composting are the cost for site preparation and equipment, the extended treatment period, targeting final use of the compost product, and environmental issues such as odors and dust. Some investment in equipment and site preparation is required. Composting is not a rapid stabilization process and, depending upon the technique, it could take several weeks to months to achieve stable compost. Determination of a suitable market for the compost is a critical issue to justify the extra effort in producing compost (Sikor, 1998).

2.10 CONCLUSION

Composting of organic matter is a highly economically feasible process in which organic material is converted into stable humic-like substances that can be used to improve soil properties. The bulking agents added in the agitated pile, static pile, and rotary drum composter offer physical stability to the composting biomass and preserve air spaces without periodic agitation of composting mixture. Rotary drum composter is highly efficient for fast degradation of organic materials compared to the agitated pile composting. Rotary drum composter is a highly efficient reactor and a highly capable technique in decentralized composting, which provides agitation, aeration, and mixing of the composting moisture, to produce a highly stable and homogeneous mixture of the end product (i.e., compost). Odor and leachate-related problems were not observed during the composting of waste biomass in rotary drum composting. During the composting process acidic intermediates are formed resulting pH of composting biomass is generally decrease in initial phase of composting, however, at the maturity of composting, pH of final compost stay near about pH 7. Infrared spectra of final compost generally represent aromatic structure with phenolic compounds and represent that organic materials more stabilized and became resistant to the microbial degradation.

REFERENCES

Adaskaveg, J.E., Gilbertson, R.L., and Dunlap, M.R. 1995. Effects of incubation time and temperature on in vitro selective delignifcation of silver leaf oak by *Ganoderma colossum. Applied and Environmental Microbiology* 61: 138–144.

Amir, S., Hafidi, M., Merlina, G., and Revel, J.C. 2005. Sequential extraction of heavy metals during composting of sewage sludge. *Chemosphere* 59: 801–810.

Amira, R.D., Roshanida, A.R., Rosli, M.I., Zahrah, M.F.S.F., Anuar, J.M., and Adha, C.M.N. 2011. Bioconversion of empty fruit bunches (EFB) and palm oil mill effluent (POME) into compost using *Trichoderma virens. African Journal of Biotechnology* 10: 18775–18780.

An, C.J., Huang, G.H., Li, S., Yu, H., Sun, W., and Peng, K. 2012. Influence of uric acid amendment on the in-vessel process of composting composite food waste. *Journal of Chemical Technology and Biotechnology* 87: 1558–1566.

Bosch Reig, F., GiGimeno Adelantado, J.V., and Moya Morena, M.C.M. 2002. FTIR quantitative analysis of calcium carbonate (calcite) and silica (quartz) mixtures using the constant ratio method. Application to geological samples. *Talanta* 58: 811–821.

Cao, W.P., Xu, H.C., and Zhang, H.H. 2013. Architecture and functional groups of biofilms during composting with and without inoculation. *Process Biochemistry* 48: 1222–1226.

Chang, J.I., and Hsu, T.E. 2008. Effects of compositions on food waste composting. *Bioresource Technology* 99: 8068–8074.

Chaturvedi, P.K., Seth, C.S., and Misra, V. 2006. Sorption kinetics and leachability of heavy metal from the contaminated soil amended with immobilizing agent (humus soil and hydroxyapatite). *Chemosphere* 64: 1109–1114.

Crawford, J.H. 1983. Composting of agricultural wastes—A review. *Process Biochemistry* 18: 14–18.

Cross, T. 1968. Thermophilic actinomycetes. *The Journal of applied bacteriology Society for Applied Bacteriology* 31: 36–53.

Diaz, M.J., Madejon, E., Lopez, F., Lopez, R., and Cabrera, F. 2002. Optimization of the rate vinasse/grape marc for co-composting process. *Process Biochemistry* 37: 1143–1150.

Dix, N.J., and Webster, J. 1995. *Fungal Ecology*. Chapman & Hall, Cambridge.

Eiland, F., Klamer, M., Lind, A.M., Leth, M., and Baath, E. 2001. Influence of initial C/N ratio on chemical and microbial composition during long term composting of straw. *Microbial Ecology* 41: 272–280.

Farnet, AM., Qasemian, L., Peter-Valence, F., Ruaudel, F., Savoie, J.M., and Ferre, E. 2013. Capacity for colonization and degradation of horse manure and wheat-straw-based compost by different strains of *Agaricus subrufescens* during the first two weeks of cultivation. *Bioresource Technology* 131: 266–273.

Fernandez, F.J., Arias, V.S., Rodriguez, L., and Villasenor, J. 2010. Feasibility of composting combinations of sewage sludge, olive mill waste and winery waste in a rotary drum reactor. *Waste Management* 30: 1948–1956.

Ge, J., Huang, G., Huang, J., Zeng, J., and Han, L. 2015. Mechanism and kinetics of organic matter degradation based on particle structure variation during pig manure aerobic composting. *Journal of Hazardous Materials* 292: 19–26.

Ghazifard, A., Kasra-Kermanshahi, R., and Far, Z.E. 2001. Identification of thermophilic and mesophilic bacteria and fungi in Esfahan (Iran) municipal solid waste compost. *Waste Management and Research* 19: 257–261.

Godden, B., Ball, A.S., Helvenstein, P., McCarthy, A.J., and Penninckx, M.J. 1992. Towards elucidation of the lignin degradation pathway in actinomycetes. *Journal of General Microbiology* 138: 2441–2448.

Grube, M., Lin, J.G., Lee, P.H., and Kokorevicha, S. 2006. Evaluation of sewage sludge based compost by FT-IR spectroscopy. *Geoderma* 130: 324–333.

Guo, R., Li, G.X., Jiang, T., Shuchardt, F., Chen, T.B., Zhao, Y.Q., and Shen, Y.J. 2012a. Effect of aeration rate, C/N ratio and moisture content on the stability and maturity of compost. *Bioresource Technology* 112: 171–178.

Guo, X., Gu, J., Gao, H., Qin, Q., Chen, Z., Shao, L., Chen, L., Li, H., Zhang, W., Chen, S., and Liu, J. 2012b. Effects of Cu on metabolisms and enzyme activities of microbial communities in the process of composting. *Bioresource Technology* 108: 140–148.

Gupta, G., Srivastava, S., Khare, S.K., and Prakash, V. 2014. Extremophiles: An overview of microorganism from extreme environment. *International Journal of Agriculture, Environment and Biotechnology* 7(2): 371–380.

Haberhauer, G., Rafferty, B., Strebl, F., and Gerzabek, M.H., 1998. Comparison of the composition of forest soil litter derived from three different sites at various decompositional stages using FTIR spectroscopy. *Geoderma* 83: 331–342.

Hachicha, R., Rekik, O., Hachicha, S., Ferchichi, M., Woodward, S., Moncef, N., Cegarra, J., and Mechichi, T. 2012. Co-composting of spent coffee ground with olive mill wastewater sludge and poultry manure and effect of Trametes versicolor inoculation on the compost maturity. *Chemosphere* 88: 677–682.

Han, W., Clarke, W., and Pratt, S. 2014. Composting of waste algae: A review. *Waste Management* 34(7): 1148–1155.

Haroun, M., Idris, A., and Omar, S. 2009. Analysis of heavy metals during composting of the tannery sludge using physicochemical and spectroscopic techniques. *Journal of Hazardous Materials* 65: 111–119.

Harun, N.A.F., Baharuddin, A.S., Zainudin, M.H.M., Bahrin, E.K., Naim, M.N., and Zakaria, R. 2012. Cellulose production from treated oil palm empty fruit bunch degradation by locally isolated Thermobifidafusca. *Bio Resources* 8: 676–687.

Haug, R.T. 1993. *The Practical Handbook of Compost Engineering*. Lewis Publishers, Boca Raton, FL.

Hsu, J.H., and Lo, S.L. 2001. Effect of composting on characterization and leaching of copper, manganese, and zinc from swine manure, *Environmental Pollution* 114: 119–127.

Huang, G.F., Wong, J.W.C., We, Q.T., and Nagar, B.B. 2004. Effect of C/N on composting of pig manure with sawdust. *Waste Management* 24: 805–813.

Huang, G.F., Wu, Q.T., Wong, J.W.C., and Nagar, B.B., 2006. Transformation of organic matter during co-composting of pig manure with sawdust. *Bioresource Technology* 97: 1834–1842.

Jeris, J.S., and Regan, R.W. 1973. Controlling environmental parameters for optimum composting. *Composting Science and Utilization* 14: 10–15.

Jørgensen, H., Errikson, T., Børjesson, J., Tjerneld, F., and Olsson, L. 2003. Purification and characterization of five cellulases and one xylanases from Penicillium brasilianum IBT 20888. *Enzyme and Microbial Technology* 32: 851–861.

Kalamdhad, A.S. 2010. *High Rate Composting of Municipal Solid Waste-An Option for Decentralized Composting*, Lambert Academic Publishing GmbH & Co. KG, Saarbrucken.

Kalamdhad, A.S., Singh, Y.K., Ali, M., Khwairakpam, M., and Kazmi, A.A. 2009. Rotary drum composting of vegetable waste and tree leaves. *Bioresource Technology* 100: 6442–6450.

Kazeem, M.O., Shah, U.K.M., and Baharuddin, A.S. 2017. Prospecting agro-waste cocktail: Supplementation for cellulase production by a newly isolated thermophilic *B. licheniformis* 2D55. *Applied Biochemistry and Biotechnology* 182: 1318–1340.

Kumar, M., Ou Yan L., and Lin J.G. 2010. Co-composting of green waste and food waste at low C/N ratio. *Waste Management* 30: 602–609.

Kutzner, H.J. 2001. Microbiology of composting. *Biotechnology* 11: 35–100.

Leconte, M.C., Mazzarino, M.J., Satti, P., Iglesias, M.C., and Laos, F. 2009. Co-composting rice hulls and/or sawdust with poultry manure in NE Argentina. *Waste Management* 29: 2446–2453.

Liu, J., Xu, X.-h., Li, H.-t., and Xu, Y. 2011. Effect of microbiological inocula on chemical and physical properties and microbial community of cow manure compost. *Biomass and Bioenergy* 35: 3433–3439.

Mouchacca, J. 1997. Thermophilic fungi: Biodiversity and taxonomic status. *Cryptogamie Mycologie* 18: 19–69.

Nakasaki, K., Sasaki, M., Shoda, M., and Kubota, H. 1985. Characteristic of mesophilic bacteria isolates isolated during thermophilic composting of sewage sludge. *Applied and Environmental Microbiology* 49: 42–45.

Nakasaki, K., Araya, S., and Mimoto, H. 2013. Inoculation of *Pichia kudriavzevii* RB1 degrades the organic acids present in raw compost material and accelerates composting. *Bioresource Technology* 144: 521–528.

Ogunwande, G.A., Osunade, K.O., Adekalu, K.O., and Ogunjimi, L.A.O. 2008. Nitrogen loss in chicken litter compost as affected by carbon to nitrogen ratio and turning frequency. *Bioresource Technology* 99: 7495–7503.

Oliveira, T.B., Lopes, V.C.P., Barbosa, F.N., Ferro, M., Meirelles, L.A., Sette, L.D., Gomes, E., and Rodrigues, A. 2016. Fungal communities in press mud composting harbours beneficial and detrimental fungi for human welfare. *Microbiology* 162: 1147–1156.

Ouatmane, A., Provenzano, M.R., Hafidi, M., and Senesi, N. 2000. Compost maturity assessment using calorimetry, spectroscopy and chemical analysis. *Compost Science and Utilization* 8: 124–134.

Pérez, J., Muñoz-Dorado, J., De-la-Rubia, T., and Martínez, J. 2002. Biodegradation and biological treatments of cellulose, hemicellulose and lignin: An overview. *International Microbiology* 5: 53–63.

Piceno, Y.M., Pecora-Black, G., Kramer, S., Roy, M., Reid, F.C., and Dubinsky, E.A. 2017. Bacterial community structure transformed after thermophilically composting human waste in Haiti. *PLoS One* 12(6): 0177626.

Provenzano, M.R., Malerba, A.D., Pezzolla, D., and Gigliotti, G. 2014. Chemical and spectroscopic characterization of organic matter during the anaerobic digestion and successive composting of pig slurry. *Waste Management* 34: 653–660.

Rabinovich, M.L., Bolobova, A.V., and Vasil'chenko, L.G. 2004. Fungal decomposition of natural aromatic structures and xenobiotics: A review. *Applied Biochemistry and Microbiology* 40: 1–17.

Rastogi, G., Muppidi, G.L., Gurram, R.N., Adhikari, A., Bischoff, K.M., Hughes, S.R., and Sani, R.K. 2009. Isolation and characterization of cellulose-degrading bacteria from the deep subsurface of the Homestake gold mine, Lead, South Dakota, USA. *Journal of Industrial Microbiology and Biotechnology* 36: 585–598.

Rastogi, G., Bhalla, A., Adhikari, A., Bischoff, K.M., Hughes, S.R., Christopher, L.P., and Sani, R.K. 2010. Characterization of thermostable cellulases produced by Bacillus and Geobacillus strains. *Bioresource Technology* 101: 8798–8806.

Ridha, H., Olfa, R., Salma, H., Mounir, F., Steve, W., Nasri, M., Juan, C., and Tahar, M. 2012. Co-composting of spent coffee ground with olive mill wastewater sludge and poultry manure and effect of *Trametes versicolor* inoculation on the compost maturity. *Chemosphere* 88: 677–682.

Rodriguez, M.E., Narros, G.A., and Molleda, J.A. 1995. Wastes of multilayer containers as substrate in composting processes. *Journal of the Air and Waste Management Association* 45: 156–160.

Rodriguez, L., Cerrillo, M.I., Garcia-Albiach, V., and Villasenor, J. 2012. Domestic sewage sludge composting in a rotary drum reactor: Optimizing the thermophilic stage. *Journal of Environmental Management* 108: 284–291.

Ruggieri, L., Gea, T., Artola, A., and Sanchez, A. 2009. Air filled porosity measurements by air pycnometry in the composting process: A review and a correlation analysis. *Bioresource Technology* 100: 2655–2666.

Sánchez, C. 2009. Lignocellulosic residues: Biodegradation and bioconversion by fungi. *Biotechnology Advances* 27: 185–194.

Shukla, O.P., Rai, U.N., and Dubey, S. 2009. Involvement and interaction of microbial communities in the transformation and stabilization of chromium during the composting of tannery effluent treated biomass of *Vallisneria spiralis* L. *Bioresource Technology* 100: 2198–2203.

Sikor, L.J. 1998. Benefits and drawbacks to composting organic by-products, in S. Brown, J.S. Angle and L. Jacobs (eds.) *Beneficial Co-Utilization of Agricultural, Municipal and Industrial By-products*, pp. 69–77, Kluwer Academic Publishers, Norwell, MA.

Singh, J., and Kalamdhad, A.S. 2011. Effects of heavy metals on soil, plants, human health and aquatic life. *International Journal of Research in Chemistry and Environment* 1(2): 15–21.

Singh, J., and Kalamdhad, A.S. 2012. Concentration and speciation of heavy metals during water hyacinth composting. *Bioresource Technology* 124: 169–179.

Singh, J., and Kalamdhad, A.S. 2016. Effect of lime on speciation of heavy metals during composting of water hyacinth. *Frontiers of Environmental Science and Engineering* 10(1): 93–102.

Singh, G., Bhalla, A., Kaur, P., Capalash, N., and Sharma, P. 2011. Laccase from prokaryotes: A new source for an old enzyme. *Reviews in Environmental Science and Biotechnology* 10(4): 309–326.

Smidt, E., Lechner, P., Schwanninger, M., Haberhauer, G., and Gerzabek, M.H., 2002. Characterization of waste organic matter by FT-IR spectroscopy—Application in waste science. *Applied Spectroscopy* 56: 1170–1175.

Smith, B., 1999. *Infrared Spectral Interpretation*. CRC Press, London.

Strom, P.F. 1985. Effect of temperature on bacterial species diversity in thermophilic solid-waste composting. *Applied and Environmental Microbiology* 50: 899–905.

Sutripta, S., Rajdeep, B., Sunanda, C., Pradeep, D., Sandipan, G., and Subrata, P. 2010. Effectiveness of inoculation with isolated Geobacillus strains in the thermophilic stage of vegetable waste composting. *Bioresource Technology* 101: 2892–2895.

Tan, K.H., 1993. *Humus and Humic Acids. Principles of Soil Chemistry: Colloidal Chemistry of Organic Soil Constituents*, pp. 79–127. Marcel Dekker UNC, New York.

Tandy, S., Healey, J.R., Nason, M.A., Williamson, J.C., Jones, D.L., and Thain, S.C. 2010. FTIR as an alternative method for measuring chemical properties during composting. *Bioresource Technology* 101(14): 5431–5436.

Tang, J.C., Wang, M., Zhou, Q.X., and Nagata, S. 2011. Improved composting of Undariapinnatifida seaweed by inoculation with *Halomonas* and *Gracilibacillus* sp. isolated from marine environments. *Bioresource Technology* 102: 2925–2930.

Tseng, D.Y., Vir, R., Traina, S.J., and Chalmers, J.J. 1996. A Fourier-transform infrared spectroscopic analysis of organic matter degradation in a bench-scale solid substrate fermentation (composting) system. *Biotechnology and Bioengineering* 52: 661–671.

Tuomela, M., Vikman, M., Hatakka, A., and Itavaara, M. 2000. Biodegradation of lignin in a compost environment: A review. *Bioresource Technology* 72: 169–183.

Vishan, I., Laha, A., and Kalamdhad, A. 2017a. Biosorption of Pb (II) by *Bacillus badius* AK strain originating from rotary drum compost of water hyacinth. *Water Science and Technology* 75(5): 1071–1083.

Vishan, I., Sivaprakasam, S.K., and Kalamdhad, A. 2017b. Biosorption of lead using *Bacillus badius* AK strain isolated from compost of green waste (water hyacinth). *Environmental Technology* 38: 13–14.

Waksman, S.A., Cordon, T.C., and Hulpoi, N. 1939. Influence of temperature upon the microbiological population and decomposition processes in composts of stable manure. *Soil Science* 47: 83–114.

Xing-Na, W., Ren-Xiang, T., and Ji-Kai, L. 2005. Xylactam, a new nitrogen-containing compound from the fruiting bodies of ascomycete Xylaria euglossa. *The Journal of Antibiotics* 58:268–270.

Xue, N., Wang, Q., Wu, C., Zhang, L., and Xie, W. 2010. Enhanced removal of NH3 during composting by a biotrickling filter inoculated with nitrifying bacteria. *Biochemical Engineering Journal* 51: 86–93.

Yoon, J.J., Cha, C.J., Kim, Y.S., Son, D.W., and Kim, Y.K. 2007. The brown-rot basidiomycete Fomitopsis palustris has the endo-glucanases capable of degrading microcrystalline cellulose. *Journal of Microbiology and Biotechnology* 5: 800–805.

Yu, M., Zhang, J., and Xu, Y. 2015. Fungal community dynamics and driving factors during agricultural waste composting. *Environmental Science and Pollution Research* 22: 19879. DOI: 10.1007/s11356-015-5172-5.

Zhu, N. 2007. Effect of low initial C/N ratio on aerobic composting of swine manure with rice straw. *Bioresource Technology* 98: 9–13.

Zhu, R., Wu, M., and Yang, J. 2011. Mobilities and leachabilities of heavy metals in sludge with humus soil. *Journal of Environmental Sciences* 23: 247–254.

3 Analysis of Compost Samples

3.1 PHYSICAL ANALYSIS

3.1.1 Moisture

Moisture contents. Moisture content (MC) (wet basis) was measured by drying at 105°C for approximately 24 hours or at constant weight.

3.1.2 Water-Holding Capacity (WHC)

A wet sample of known initial MC was weighed (W_i) and placed in a beaker. After soaking in water for 1–2 days and draining excess water through Whatman 2 No. filter paper, the saturated sample was weighed again (W_s). The amount of water retained by dry sample was calculated as the WHC. The WHC (g water/g dry material) can be calculated as follows (Ahn et al., 2008):

$$\mathrm{WHC} = \frac{\left[(W_\mathrm{s} - W_\mathrm{i}) + \mathrm{MC} \times W_\mathrm{i}\right]}{\left[(1 - \mathrm{MC}) \times W_\mathrm{i}\right]}, \tag{3.1}$$

where W_i is the initial weight of sample (g), W_s is the final weight of sample (g), and MC is the initial MC of sample (decimal).

3.1.3 Water-Absorption Capacity (WAC)

Adhikari et al. (2009) described a method to determine the WAC value of a compost sample. The compost sample is soaked in distilled water for 24 hours and then the wet sample is dried at 105°C for 24 hours after the gravitational water is drained off under cover. The total water absorbed is the difference between the weight of a sample totally immersed in water and its dry weight:

$$\mathrm{WAC}(\%) = \frac{100(W_\mathrm{AB} - W_\mathrm{DRY})}{W_\mathrm{DRY}}, \tag{3.2}$$

where W_{AB} is the weight of the soaked sample (g) and W_{DRY} is the weight of dried sample (g).

3.1.4 Bulk Density (BD)

BD was analyzed using an approximately 10 L volume container. The container was filled with solid material, and then the material was slightly compressed to confirm nonexistence of large void spaces. The BD is the ratio of the weight of the material to the volume of material in the container (Khater, 2012).

3.1.5 Compost Porosity

Compost porosity (ε_a) can be measured using the known density of water (ρ_w; 1000 kg m^{-3}) and the estimated densities of organic matter (OM) (ρ_{om}; 1600 kg m^{-3}), and ash (ρ_{ash}; 2500 kg m^{-3}), as well as the MC and bulk densities of the compost sample. Thus, the compost porosity can be calculated as follows (Khater, 2012):

$$\varepsilon_a = 1 - \rho_{wb}\left(\frac{MC}{\rho_w} + \frac{DM \cdot OM}{\rho_{om}} + \frac{DM \cdot (1 - OM)}{\rho_{ash}}\right) \times 100, \quad (3.3)$$

where ε_a is the compost porosity (%), ρ_{wb} is the wet BD (kg m^{-3}), ρ_w is the density of water (kg m^{-3}), ρ_{om} is the density of OM (kg m^{-3}), ρ_{ash} is the density of ash (kg m^{-3}), MC is the moisture content (decimal), DM is the dry matter (decimal), and OM is the organic matter (decimal).

3.1.6 Free Air Space (FAS)

FAS can be calculated as follows (Epstein, 1997):

$$FAS(\%) = 100\left(1 - \frac{BD}{SG}\right), \quad (3.4)$$

where BD is the BD (g cm^{-3}) and SG is the specific gravity (g cm^{-3}) of the raw material or the composting mixture.

3.1.7 Temperature

Temperature was measured at a 6 hour interval by a digital thermometer throughout the composting period within the agitated pile and composter (horizontal reactor).

3.2 CHEMICAL ANALYSIS

Figure 3.1 shows a flow chart of physico-chemical analysis of compost samples.

3.2.1 Volatile Solid (VS) and Ash Content

VS and ash contents were also measured according to Singh and Kalamdhad (2012). Initial weight of the crucible was taken as W_1 g. Take 10 ± 0.1 g of ground sample (screened through 0.22 mm sieve) in a crucible and keep it in a muffle furnace at a temperature of 550°C–600°C for 2 hours. After 2 hours, remove the crucible from the muffle furnace and keep it in the desiccator for ½ hour for cooling and then measure the final weight of the crucible with the sample as W_2 g. VS content of the sample can be calculated as follows:

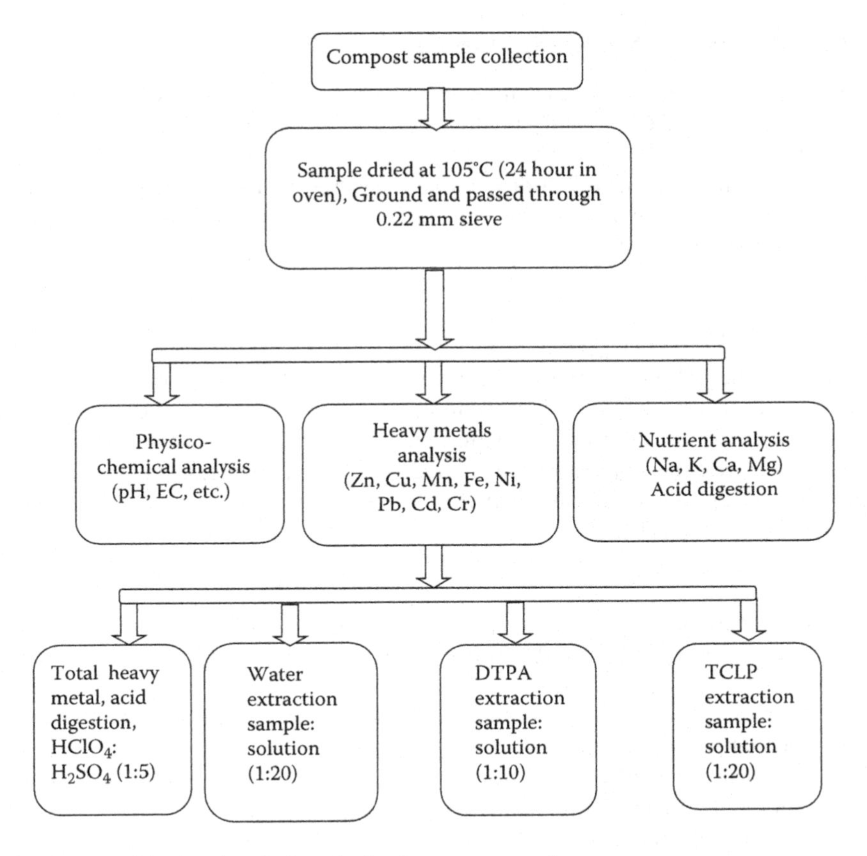

FIGURE 3.1 Physicochemical analysis of compost samples.

$$\text{VS }(\%) = 100\left(5 - \frac{W_2 - W_1}{5}\right) \times 100, \tag{3.5}$$

$$\text{Ash content}(\%) = 100 - \text{VS}. \tag{3.6}$$

The biodegradability coefficient (K_b) is calculated as follows (Yadav and Garg 2009):

$$K_b = \left(\frac{(\text{OM}_i - \text{OM}_f)\ 100}{\text{OM}_i\ (100 - \text{OM}_f)}\right), \tag{3.7}$$

where OM_i and OM_f are the initial and final OM contents of the composting process, respectively.

3.2.2 Analysis of Na, K, Mg, and Ca

- Take 0.2 g of the compost sample in a block digestion system with 10 mL of H_2SO_4 and $HClO_4$ (5:1) mixture for 2 hours at 300°C (Singh and Kalamdhad, 2013).
- Na, K, and Ca concentrations can be measured by the Flame photometer, whereas Mg concentration can be measured by atomic absorption spectrometer (AAS).

3.2.3 The pH

Principle. The pH value, which is a measure of the hydrogen (or hydroxyl)-ion activity of the compost–water system, indicates whether the compost is acidic, neutral, or alkaline in reaction. pH stands for potential hydrogen and it measures the concentration of hydrogen ion in water. The pH is calculated by taking the negative logarithm of the hydrogen ion concentration as follows: $pH = -\log [H^+]$.

Procedure

- Take approximately 10 ± 0.1 g of the sample in powder form (passed through 0.22 mm sieve) in a 250 mL reagent bottle and mix with 100 mL of distilled water.
- Keep this mixture in a horizontal water bath shaker for 2 hours at 100 rpm.
- Then let the solids settle for about 30 minutes.
- Standardize the pH meter with immersing the electrode is used and after calibrated with the buffer solutions of pH 4.0 and 7.0.
- Rinse the electrode two to three times with distilled water.
- The pH value of the extracted solution can be measured by a pH meter.

3.2.4 Electrical Conductivity (EC)

Principle. EC is the measure of the ability of the substance/solution to allow the flow of an electric current (this electric current is conducted by ions existing in the extracted solution).

Procedure

- Take approximately 10 ± 0.1 g of the sample in powder form (passed through 0.22 mm sieve) in a 250 mL conical flask and mix with 100 mL of distilled water.
- Keep this mixture in a horizontal shaker for 2 hours at 100 rpm.
- Then let the solids in the mixture settle for 30 minutes and separate the extracted solution through Whatman 42 No. filter paper.
- Rinse the electrode two to three times with distilled water.
- EC of the extracted solution can be measured by an electrical conductivity meter in dS/m.

3.2.5 Total Nitrogen

Principle. The determination of the total nitrogen in the compost sample helps to evaluate the fertility status of the soil (Tiquia and Tam, 2000). Nitrogen in soil and sediment is present mainly in the organic form together with small amounts of ammonia and nitrate. Kjeldahl's method measures only organic and ammonium forms; nitrate is excluded.

The total nitrogen composite is the sum of total Kjeldahl nitrogen (TKN), nitrate nitrogen (NO_3-N), ammonia nitrogen, and organic nitrogen present in the sample (excluding nitrite nitrogen and nitrate nitrogen):

$$TN = TKN + NO_3\text{-}N \tag{3.8}$$

$$TKN = \text{organic N} + NH_4\text{-}N \tag{3.9}$$

In the digestion step, the organic nitrogen as well as ammonia nitrogen is converted to an ammonium sulfate at 360°C–410°C in the presence of sulfuric acid, potassium sulfate, and cupric sulfate (catalysts). An excess of alkali was then added and the liberated ammonia was distilled into a boric acid solution (mixed with indicator). TKN is determined titrimetrically with standard solutions of sulfuric acid or hydrochloric acid.

Procedure

- For TKN analysis, take 0.2 g of ground sample in a digestion tube, add 3–4 g of the catalyst mixture (potassium sulfate and cupric sulfate in 5:1 ratio), and finally add 10 mL of conc. Keep sulfuric acid in the digestion flask at 350°C–420°C.
- After the digestion step, prepare a solution up to 100 mL with distilled water and then take 10 mL of the diluted sample. Add 20 mL of 40% NaOH and keep this mixture in a distillation tube.
- Then after starting the distillation process, the tip of the condenser should be immersed in a 25 mL boric acid solution (mixed with indicator) in a 250 mL conical flask. Collect approximately 150 mL of the distillated solution.
- Titrate the collected solution with the standardized H_2SO_4 (0.02 N) or HCl (0.1 N).
- Thus, the appearance of a purple color indicates the end point of titrated solution.

$$TKN(\%) = \left[\frac{(A - B) \times N \times 14}{Wt.}\right] \times 10, \tag{3.10}$$

where A indicates mL of standard H_2SO_4 (or HCl) used for sample, B indicates mL of standard H_2SO_4 (or HCl) used for blank, N is normality of standard H_2SO_4 (or HCl), and Wt. is the weight of the sample (g).

3.2.6 Nitrate (NO_3-N)

Principle. An ultraviolet (UV) technique suitable for screening uncontaminated water was used to measure the absorbance of nitrate at 220 nm (low in OM). Since the other organic matter may absorb at 275 nm and nitrate does not absorb at 275 nm, a second measurement was made at 275 nm to find out the correct value of the nitrate. The calibration curve of nitrate follows Beer's law up to maximum nitrate concentration of 11 mg NO_3-N/L

Procedure

- Take approximately 3 g of the sample in powder form in a conical flask and add 30 mL of 2M KCl to the compost sample.
- Keep this mixture in a horizontal water bath shaker for 2 hours at 100 rpm.
- After shaking, separate the extracted solution from the solid residue using Whatman 42 No. filter paper.
- Take 50 mL of the extracted solution and add 1.0 mL of 1.0 N HCl solution and then mix it properly.
- The analysis for NO_3-N can be performed by using UV–visible spectrophotometry

Preparation of standard curve. Prepare nitrate calibration standards in the range of 0–7 mg NO_3-N/L by diluting to 50 mL the following volumes of intermediate nitrate solution with distilled water: 0, 1, 2, 4, 7, 10, 15, 20, 25, 30, and 35–. Treat the nitrate standards in the same manner as samples. Set the spectrophotometer to zero absorbance with distilled water. Calculate the absorbance at 220 nm wavelength to obtain NO_3-N reading and at 275 nm wavelength to determine the interference caused due to dissolved OM.

Calculation

- Reduce the absorbance reading two times at 275 nm wavelength to determine the extracted sample and standards and at 220 nm wavelength to find the absorbance measured due to NO_3-N.
- Make a standard curve by using absorbance of nitrate against its known concentration.
- Calculate nitrate concentration directly from the standard curve.

3.2.7 Ammonical Nitrogen (NH_4-N).

Principle. This mixture catalyzed by sodium nitroprusside—a strongly blue compound known as indophenol—is formed by the reaction of ammonia, hypochlorite, and phenol (APHA, 2005).

Procedure

- Take approximately 3 g of the sample in powder form in a conical flask and add 30 mL of 2M KCl to the compost sample.
- Keep the conical flask in a horizontal water bath shaker for 2 hours at 100 rpm.

- After shaking, separate extracted solution by a filter.
- The analysis of NH_4^+-N can be performed by using the Phenate method.

Phenate method

- Take 25 mL of supernatant in a 50 mL Erlenmeyer flask and add 1 mL of phenol solution, followed by 1 mL of sodium nitroprusside solution and 2.5 mL of oxidizing solution. Then, mix it properly with the compost sample and reagents.
- Cover the mixture with plastic wrapping or wax film. Allow color to develop in restrained light for at least 1 hour at room temperature of 22°C–27°C. The stability of color remains for 24 hours.
- Measure the absorbance of the color developed at 640 nm.
- Prepare a blank and standard solution by diluting stock ammonia solution as the sample concentration range. Treat all the standards in the same manner as samples.

Calculations

Prepare a standard curve by plotting absorbance readings of standards against ammonia concentrations of standards. Calculate the ammonia concentration from the standard calibration curve.

3.2.8 Total Phosphorus

Principle. The organic phosphorus is converted into inorganic form after digestion with conc. Nitric acid. Phosphorus in the form of phosphate reacts with ammonium molybdate to form phosphomolybdic acid, which gives a blue color complex in the presence of stannous ions and can be determined spectrophotometrically at 690 nm.

Procedure

- Take 0.2 g of oven dried (105°C) sample thoroughly ground and sieved through a 0.2 mm sieve in a digestion tube and add 10 mL of mixed acid solution (perchloric acid and sulfuric acid in the ratio of 1:5).
- Digest it at 300°C.
- Cool down the digested sample and add distilled water until its volume is 100 mL.
- Take 50 mL of the diluted sample in a beaker and add 0.05 mL phenolphthalein indicator solution. If a red color develops, add 1 mL of H_2SO_4 solution and either 0.4 g solid ammonium persulfate or 0.5 g solid sodium persulfate.
- Boil gently using a hot plate for about 30 minutes until the final volume is reduced to 10 mL.
- Cool down and dilute with 30 mL distilled water and add 0.05 mL (1 drop) phenolphthalein indicator solution to diluted sample and neutralize with NaOH. After a few minutes, a very faint pink color develops.

- Add distilled water until the final volume is 100 mL and then add 1 drop of phenolphthalein indicator to the solution. If the sample turns pink, add strong acid solution dropwise until the sample turns colorless. If more than five drops are necessary, take a smaller amount of the sample and dilute it to 100 mL with distilled water after first discharging the pink color with acid.

Color development

- Add 4 mL molybdate reagent and 10 drops of stannous chloride reagent. Then mix it thoroughly.
- A blue color will appear. Wait for 10 minutes and record absorbance at 690 nm before 12 minutes using a spectrophotometer. Carry out a blank determination using distilled water.
- Determine the concentration from the standard calibration curve (preferably 0.0, 0.1, 0.2, 0.3, 0.4, and 0.5 mg of P/L) for phosphate prepared from appropriate dilution of the working standard phosphate solution.

3.2.9 Available Phosphorus

Procedure

- Take 0.2 g of oven dried (105°C) sample thoroughly ground and sieved through a 0.2 mm sieve in a digestion tube and add 10 mL of mixed acid solution (perchloric acid and sulfuric acid in the ratio of 1:5).
- Digest it at 300°C.
- Cool down the digested sample and then add distilled water until its volume is 100 mL.

Color development

- Add 4 mL molybdate reagent and 10 drops of stannous chloride reagent and mix it thoroughly.
- A blue color will appear. Wait for 10 minutes and record absorbance at 690 nm before 12 minutes using a spectrophotometer. Carry out a blank determination using distilled water.
- Determine the concentration from the standard calibration curve (preferably 0.0, 0.1, 0.2, 0.3, 0.4, and 0.5 mg of P/L) for phosphate prepared from appropriate dilution of the working standard phosphate solution.

3.2.10 Chemical Oxygen Demand (COD)

Principle. Most types of the OMs are oxidized when boiled with a mixture of potassium dichromate and sulfuric acid and produce carbon dioxide and water. A sample is refluxed with a known amount of potassium dichromate and sulfuric acid medium, and the excess dichromate is titrated with ferrous ammonium sulfate. The amount of dichromate consumed during titration is proportional to the oxygen required to oxidize the oxidizable organic matter.

Procedure

- Take approximately 20 ± 0.1 g of fresh compost sample in a conical flask and dissolve this sample with 200 mL of distilled water.
- Keep this mixture in a horizontal water bath shaker for 2 hours at 100 rpm.
- Then let the solids settle for 30 minutes.
- Take supernatant after 10 minutes centrifugation at 6000–7000 rpm.
- The collected supernatant of samples can be used for COD analysis by a closed reflux method.
- Take 2.5 mL of supernatant and add 1.5 mL of $K_2Cr_2O_7$ and 3.5 mL of sulfuric acid to a COD vial.
- Keep this mixture in a COD digester for 2 hours at 150°C.
- Cool down the digested sample at room temperature and then titrate with standard ferrous ammonium sulfate using ferroin indicator. When the color of solution changes sharply from green blue to wine red, the end point of the titration has been reached.
- Reflux a reagent blank under identical conditions simultaneously with the sample.

Calculation

$$\text{Molarity of FAS solution} = \frac{\text{Volume } 0.01667\text{M } K_2Cr_2O_7 \text{ solution titrated, mL}}{\text{Volume of FAS used in titration, mL}} \times 0.25, \quad (3.11)$$

$$\text{COD as mg } \frac{O_2}{L} = \frac{(V_1 - V_2) \times M \times 8000}{V_0}, \quad (3.12)$$

where V_1 is the volume of Fe $(NH_4)_2(SO_4)_2$ required for titration against the blank in mL, V_2 is the volume required for titration against the sample in mL, M is the molarity of Fe $(NH_4)_2(SO_4)_2$, and V_0 is the volume of sample taken for testing in mL.

3.2.11 Total Heavy Metal Content

Approximately 0.2 g sample is digested with 10 mL of H_2SO_4 and $HClO_4$ (5:1) mixture in a block digestion system for 2 hours at 300°C or using a hot plate. The digested sample can be used to measure the total concentration of heavy metals through AAS or other instruments.

3.2.12 Extraction of Water-Soluble Heavy Metals

- Take approximately 2.5 g of the compost sample extracted with 50 mL of distilled water (sample: solution ratio = 1:20) at room temperature using a bath shaker for 2 hours at 100 rpm (Singh and Kalamdhad, 2013).
- Keep the mixture for 5 minutes at 10,000 rpm, and then separate the mixture through Whatman No. 42 filter paper and store in a plastic reagent bottle at 4°C for heavy metal analysis by AAS.

3.2.13 Extraction of Heavy Metals with Diethylene Triamine Penta-Acetic Acid (DTPA)

- Take approximately 4 g of the sample in powder form and add 40 mL of 0.005 M DTPA, 0.01 M $CaCl_2$, and 0.1 M (triethanolamine) to the sample with pH 7.3 shaking at 100 rpm (Guan et al., 2011).
- The concentration of heavy metals in the extracted solution can be measured by using AAS. DTPA extraction efficiency of heavy metals can be calculated by using the following equation (Chen, 2010):

$$\text{Extraction effciency}(\%) = \frac{C_{\text{DTPA}}}{C_{\text{Total}}} \times 100, \tag{3.13}$$

- where C_{DTPA} is the concentration of DTPA extractable heavy metal and C_{Total} is the concentration of total heavy metals.
- The remaining steps are similar to the extraction of water-soluble heavy metals.

3.2.14 Leachability of Heavy Metals

Principle. The standard toxicity characteristic leaching procedure (TCLP) method (US EPA, 1992) was performed to the compost samples in order to determine the potential leachability of heavy metals from the compost sample.

Procedure

- Take approximately 5 g compost sample (<9.5 mm of size) in a 125 mL reagent bottle and mix with 100 mL of acetic acid at pH 4.93 ± 0.05 (adjust pH using 1 N NaOH) (sample: solution ratio = 1:20) at room temperature using a bath shaker for 18 hours at 30 ± 2 rpm.
- The remaining steps are similar to the extraction of water-soluble heavy metals.

3.2.15 Chemical Speciation of Heavy Metals

A sequential extraction of the heavy metals in the compost samples can be carried out by using method of Tessier et al. (1979). These authors reported the following five forms of fractions: (i) exchangeable fraction (F1), (ii) carbonate fraction (F2),

TABLE 3.1
The Extractants and Extraction Environmental of Heavy-Metal Forms with Sequential Extraction

S. No.	Speciation	Extractants	Extraction Environments
1	Exchangeable fraction (F1)	• 8 mL of 1.0 M $MgCl_2$ (pH=7)	With agitation at 220 rpm for 1 hour at 25°C
2	Carbonate fraction (F2)	• 8 mL 1.0 M NaOAc (pH=5, adjusted with conc. HOAc).	With continuous agitation for 5 hours at 25°C
3	Reducible or FeMn bound fraction (F3)	• 20 mL 0.04 M $NH_2OH.HCl$ in 25% HOAc (v/v).	Kept in water bath shaker for 6 hours at 96°C with occasional agitation
4	Oxidizable/ organic matter bound fraction (F4)	• 3 mL 0.02 M HNO_3 and 5 mL 30% H_2O_2 (pH = 2, adjusted with conc. HNO_3). • After 2 hours, 3 mL 30% H_2O_2 was added. • After cooling, 5 mL of 3.2 M NH_4OAc in 20% (v/v) HNO_3 was added.	Mixture was heated at 85°C for 2 hours. Mixture was heated at 85°C for 3 hours with occasional agitation. Agitated for 0.5 hours at 25°C
5	Residual fraction (F5)	• 10 mL of H_2SO_4 and $HClO_4$ (5:1) mixture	Heated at 300°C 2 hours

(iii) reducible fraction (F3), (iv) organically bound fraction (F4), and (v) residual fraction (F5). The procedure for sequential extraction is presented in Table 3.1.

The extraction can be carried out with an initial mass of 1.0 g oven dried sample in polypropylene centrifuge tubes of 50 mL capacity. After each successive extraction, the supernatant liquid is separated with a pipette after centrifugation at 10,000 rpm for 5 minutes and diluted for heavy metal analysis. The residue is washed (except residual fraction) with 20 mL of Milli Q water by shaking for 15 minutes followed by centrifugation without loss of solids. The extracts are stored in plastic reagent bottles for heavy metal analysis. The concentration of heavy metals in each extract is determined by AAS. The bioavailability factor (BF) can be calculated as follows (Liu et al., 2007):

$$BF = \frac{F1 + F2 + F3 + F4}{F1 + F2 + F3 + F4 + F5}. \tag{3.14}$$

3.3 BIOLOGICAL ANALYSIS

3.3.1 CO_2 Evaluation by Soda-lime Method

Principle. Carbon dioxide evolution is the direct method to measure the stability of compost (Kalamdhad et al., 2008). This method measures carbon derived directly from the compost to be tested. Therefore, CO_2 evolution is directly correlated with the aerobic respiration activities.

Procedure

- Take approximately 25 ± 0.1 g of fresh compost sample in a PVC airtight container of capacity of 1 L.
- Take approximately 10 g soda lime (oven dried at 105°C) (1.5–2.0 mm mesh size) in a 100 mL beaker and place it in the PVC container. The initial weight of the soda lime taken can be measured as (W_1) g.
- Keep the container with soda-lime beaker in an incubator at 25°C.
- After a 20–24 hour incubation period, separate the soda lime and keep it in the oven in the same manner as earlier. The final weight of the soda lime can be measured as (W_2) g.
- The difference in the weight of soda lime can be used to measure the amount of CO_2 absorbed.

Calculation

$$CO_2\left(\frac{\frac{mg}{g}VS}{hour}\right)=\frac{(W_2-W_1)\times 1000}{W\times T}, \quad (3.15)$$

where W_1 is the initial weight of the soda lime (g), W_2 is the final weight of the soda lime (g), W is the weight of compost sample taken (g), and T is the time duration of incubation (hours).

3.3.2 Oxygen Uptake Rate (OUR)

Principle. OUR is a well-accepted technique for the determination of biological activity of a material (Kalamdhad et al., 2008). By using this method, the stability of compost can be evaluated by measuring the amount of easily biodegradable OM present in the sample through its carbonaceous oxygen demand.

Procedure

- Take approximately 5–8 g of the compost sample in the liquid suspension of compost with 500 mL of distilled water with the following chemical reagents: $CaCl_2$, $MgSO_4$, $FeCl_3$, and phosphate buffer at pH 7.2.
- Keep the mixture at a constant temperature by immersing the whole assembly in a water bath for temperature control of 30°C.
- During this time, the dissolved O_2 of the suspension can be continuously observed by the digital DO meter fitted with suspension.
- The dissolved oxygen consumption rate can be calculated at different time intervals and can be regarded as the OUR expressed in mg O_2/g VS/hour.

3.3.3 Biochemical Oxygen Demand (BOD)

Principle. The BOD test is an empirical bioassay-type procedure, which measures the dissolved oxygen consumed by microorganisms while assimilating and oxidizing the OM present in the aerobic conditions.

Procedure

- Take approximately 20 ± 0.1 g of fresh compost sample in a conical flask and mix with 200 mL of distilled water.
- Keep this mixture in a horizontal water bath shaker for 2 hours.
- Then let the solids settle for ½ hour.
- The collected supernatant of samples can be used for BOD analysis by a digital DO probe.

Calculation

$$\text{BOD, mg/L} = \frac{D_1 - D_2}{P} \times 100\,, \tag{3.16}$$

where D_1 is the initial DO of sample in mg/L, D_2 is the DO of sample after incubation in mg/L, and P is the sample volume (in mL) diluted to 300 mL with diluted water.

3.3.4 Coliform Analysis

The collected supernatant of 10 g fresh samples with 100 mL distilled water can be used for coliform analysis.

3.3.4.1 Total Coliform

The total coliform tests were analyzed by a multiple-tube technique using *lauryl tryptose* broth in the presumptive portion of the Standard Total Coliform Fermentation Technique by incubating inoculated tubes at 35°C ± 0.5°C for 24 ± 2 hours, and, if no gas or acidic reaction is evident, reincubate and reexamine at the end of 48 ± 3 hours. The results are estimated using the most probable number (MPN) method (APHA, 2005).

3.3.4.2 Fecal Coliform

The Fecal coliform tests are analyzed by a multiple-tube technique using EC medium by incubating inoculated tubes at 44.5°C ± 0.2°C for 24 ± 2 hours and the results are estimated using the MPN method (APHA, 2005).

3.3.5 Protein Determination by Lowry Procedure

Preparation of reagent

1. 2% Na_2CO_3 in 0.1 N NaOH
2. 1% NaK tartrate in H_2O

3. 0.5% $CuSO_4.5\ H_2O$ in H_2O
4. 48 mL of 1, 1 mL of 2, and 1 mL of 3
5. Phenol reagent: [1 part Folin-Phenol (2×) dissolved in 1 part water] (reagents 1, 2, and 3 may be stored indefinitely)

BSA standard: [Dissolve 5 mg bovine serum albumin in 5 mL of water (1 µg/µL)]. Store 1 mL aliquots in refrigerator.

Procedure

- Prepare different dilutions of BSA solutions by mixing stock BSA solution (1 mg/mL) and watering the test tubes. The final volume in each of the test tubes is 5 mL. The concentration of BSA can be measured in the range of 0.05–1 mg/mL.
- Take 0.2 mL of the protein solution from these different dilutions using pipette in various test tubes and add 2 mL of alkaline copper sulfate reagent (analytical reagent) and then mix the solutions thoroughly.
- Keep this mixture in an incubator at room temperature for 10 minutes.
- Add 0.2 mL of the reagent Folin Ciocalteau solution (reagent solutions) to each tube and keep it in the incubator for 30 minutes. Set the colorimeter to zero with blank and calculate the optical density (absorbance) at 660 nm.
- Prepare a standard calibration curve by plotting the absorbance readings of standards against protein concentration of standards.
- Measure the absorbance of an unknown sample and determine the concentration of the unknown sample from the standard calibration curve.

3.3.6 Microbial Diversity during Composting

Sample preparation for microbiological assays. Approximately 10 g of waste or compost is added into 90 mL of sterile distilled water and shake mechanically for 2 hours to perform the maximum removal of microorganisms from their organo-mineral substrates. Finally, the waste suspensions are filtered and used for microbial counts.

3.3.6.1 Total Heterotrophic Bacteria

Nutrient Agar medium (NAM) is used for the identification of culturable bacteria following the pour-plating technique for preparing serial dilutions of sample. Finally, the prepared plates were incubated in an inverted position for 24–48 hours at 25°C.

3.3.6.2 Spore Forming Bacteria

Samples are incubated at 80°C for 10 minutes to select the spores at the expense of vegetative forms. Serial dilutions of sample are plated using NAM for 24–48 hours at 25°C.

3.3.6.3 Total Fungus

The total number of viable fungus are determined by plating appropriated diluted suspensions onto Sabouraud 4% Dextrose Agar (SDA) plates containing peptone from casein 5 g, peptone from meat 5 g, glucose 40 g, and agar-agar 10 g in 1000 mL. Colonies are counted after a 72 hour incubation at 25°C.

3.3.6.4 Streptomycetes

The total counts of streptomycetes are determined by ISP medium 4. This medium can be prepared by adding 1 g K_2HPO_4, 1 g $MgSO_4.7H_2O$, 1 g NaCl, 2 g $(NH_4)_2SO_4$, 2 g $CaCO_3$, 10 g of soluble starch, 0.2 g cycloheximide, 1 mL solution of trace elements, and 20 g agar to 1 L of distilled water. The solution of trace elements is prepared by the addition of 0.1 g $FeSO_4.7H_2O$, 0.1 g $MnCl_2.4H_2O$, and 0.1 g $ZnSO_4.7H_2O$ to 1000 mL distilled water. The pH value of the medium is found to be 7.2

3.3.6.5 Actinomycetes

The total counts of actinomycetes are determined by actinomycete isolation agar, which are composed of sodium caseinate 2 g, L-asparagine 0.1 g, sodium propionate 4 g, dipotassium phosphate 0.5, magnesium sulfate 0.1, ferrous sulfate 0.001, and agar 15 g in 1000 mL distilled water. The pH value of the medium is found to be 8.1.

Calculation

$$\frac{\text{CFU}}{\text{g}} = \frac{(\text{no. of colonies} \times \text{dilution factor})}{\text{volume of culture}}. \tag{3.17}$$

Indicator organisms (fecal thermotolerant coliforms and fecal *Streptococci*) were monitored in all waste samples using the MPN method. The serial dilutions (10^1–10^8) of the samples are inoculated in a prepared medium (EC medium for fecal coliforms; azide dextrose broth for fecal *Streptococci*) and incubated at 44.5°C and 37°C for 24 hours, respectively.

3.3.6.6 *Escherichia coli*

The total counts of bacteria are determined by plating serial dilutions of sample on the MacConkey Agar medium. Plates are incubated in an inverted position for 24–48 hours at 37°C.

REFERENCES

Adhikari, B.K., Barrington, S., Martinez, J., and King, S. 2009. Effectiveness of three bulking agents for food waste composting. *Waste Management* 29(1): 197–203.

Ahn, H.K., Richard, T.L., and Glanville, T.D. 2008. Laboratory determination of compost physical parameters for modeling of airflow characteristics. *Waste Management* 28: 660–670.

American Public Health Association (APHA). 2005. *Standard Methods for the Examination of Water and Wastewater*, 21st ed. APHA, Washington, DC.

Chen, Y.X., Huang, X.D., Han, Z.Y., Huang, X., Hu, B., Shi, D.Z., and Wu, W.X. 2010. Effects of bamboo charcoal and bamboo vinegar on nitrogen conservation and heavy metals immobility during pig manure composting. *Chemosphere* 78: 1177–1181.

Epstein, E. 1997. *The Science of Composting.* Technomic Publishing Company, Lancaster, PA.

Guan, T.X., He, H.B., Zhang, X.D. and Bai, Z., 2011. Cu fractions, mobility and bioavailability in soil-wheat system after Cu-enriched livestock manure applications. *Chemosphere* 82: 215–222.

Kalamdhad, A.S., Pasha, M., and Kazmi, A.A. 2008. Stability evaluation of compost by respiration techniques in a rotary drum composter. *Resources, Conservation and Recycling* 52: 829–834.

Khater, E.S. 2012. Chemical and physical properties of compost. *Misr Journal of Agricultural Engineering* 29(4): 1–14.

Liu, Y., Ma, L., Li, Y., and Zheng, L. 2007. Evolution of heavy metal speciation during the aerobic composting process of sewage sludge. *Chemosphere* 67: 1025–1032

Singh, J., and Kalamdhad, A.S. 2012. Concentration and speciation of heavy metals during water hyacinth composting. *Bioresource Technology* 124: 169–179.

Singh, J., and Kalamdhad, A.S. 2013. Assessment of bioavailability and leachability of heavy metals during rotary drum composting of green waste (Water hyacinth). *Ecological Engineering* 52: 59–69.

Tessier, A., Campbell, P.G.C., and Bisson, M. 1979. Sequential extraction procedures for the speciation of particulate trace metals. *Analytical Chemistry* 51: 844–851.

Tiquia, S.M., and Tam, N.F.Y. 2000. Fate of nitrogen during composting of chicken litter. *Environmental Pollution* 110: 535–541.

US Environmental Protection Agency Method 1311- toxicity characteristic leaching procedure (TCLP), 35 p, July 1992.

Yadav, A., and Garg, V.K. 2009. Feasibility of nutrient recovery from industrial sludge by vermicomposting technology. *Journal of Hazardous Materials* 168: 262–268.

4 Composting and Heavy Metals

4.1 HEAVY METALS IN COMPOSTING

Composting is the best alternative for the management of sewage sludge (SS) generated from water treatment plants (Raichura and McCartney, 2006; Gea et al., 2007; Zorpas and Loizidou, 2008; Shen et al., 2012). SS is composed of a wide range of organic compounds, macronutrients, micronutrients, essential/nonessential trace metals, organic micro pollutants, and pathogens (Singh and Agrawal, 2008; Ozcan et al., 2013). SS is rich in organic matter (OM) and nutrients (N, P etc.) for plant growth. It can be used as a fertilizer in agricultural soils directly or after composting. Proper practice allows the recovery of valuable OM, N, and P to be applied as soil conditioner (Grøn, 2007; Haynes et al., 2009). SS also contains heavy metals, such as Zn, Cu, Ni, Pb, Cd, Cr, and Hg, which can affect its suitability for direct land application (Yang et al., 2014; Huang and Yuan, 2016).

The treatment of biodegradable wastes through composting process causes many physical and biochemical changes in the composting biomass, this happened due to development of microbial populations. Heavy metals are released in the process of mineralization of OM, metal solubilization affected by pH decrease, metal biosorption by the microbial biomass, or metal complexation with the humic substances (HS) formed during composting (Hsu and Lo, 2001; Zorpas et al., 2003, Haroun et al., 2009). The application of degradable wastes and its compost which may contain high concentration of heavy metals may be a danger for human health through absorption by plants from soil medium.

Heavy metals are considered nondegradable inorganic natural resources in the environment. Heavy metals with a density >5000 kg/m^3 (Talbot, 2006) are considered to be a major source of contamination of agricultural soils. Some heavy metals, such as Cu, Ni, Cd, Zn, Cr, and Pb (Karaca et al., 2010; Chiroma et al., 2012), have harmful consequences for soil microorganisms, whereas some metals (e.g., Fe, Zn, Ca, and Mg) have been reported to be of bio-importance to man and their daily uses. Some heavy metals (e.g., As, Cd, Pb, and methylated forms of Hg) have been reported to be toxic to the environment even at very low concentrations (Duruibe et al., 2007). Singh and Kalamdhad (2012) reported that the heavy-metal concentrations increase due to the degradation of OM in the composting mixture during composting of water hyacinth. Table 4.1 represents the physicochemical properties of compost prepared from different sources of materials: SS (Fang and Wong, 1999; Amir et al., 2005; He et al., 2009a,b), tannery sludge (Haroun et al., 2009), municipals solid waste (MSW) (Pichte and Anderson, 1997), swine manure (Huang et al., 2004), and water hyacinth (Prasad et al., 2013; Singh and Kalamdhad, 2013a,b). Natural humus soil is formed from decaying leaf litter that contains HS such as organic macromolecule

TABLE 4.1
Physicochemical Properties of Sewage Sludges, Tannery Sludge, MSW, Swine Manure, and Water Hyacinth

Parameter	Sewage Sludge	MSW	Tannery Sludge	Swine Manure	Water Hyacinth
pH	7.5	6.8	7.4	8.1	5.8
EC (dS/m)	1.9	4.5	9	2.9	4.9
MC (%)	82.5	–	60.6	68.3	85.9
Ash (% of dry weight)	20.13	–	65.0	–	27.4
Total carbon (% of dry weight)	38.5	42.2	20.03	36.6	40.3
TN (% of dry weight)	6.9	–	1.0	1.9	1.2
C/N	14.0	–	20.0	–	–
TP (% of dry weight)	2.0	0.4	–	0.9	0.1
Total Heavy Metals (mg/kg dry weight)					
Cu	599	236	80	320.9	39.8
Zn	728	655	200	688.7	152.0
Ni	99	28	–	–	179.8
Cd	1.2	–	8	–	43.3
Pb	191	210	10	52.2	1140
Cr	134	21	500	–	301.2
Hg	2.5	–	–	–	–
As	2.5	–	–	–	–
Mn	–	–	–	–	644.8
Fe	3768	–	1062	–	12925

humic acids, which possess multiple properties and high structural complexity (Zhu et al., 2011) Recent research has reported that natural humus soils are appropriate amendment materials for the removal of mobility and bioavailability of heavy metals (Zhu et al., 2011).

Some heavy metals, such as Zn and Cu, can be used as feed additives to reduce the length of the breeding cycle. Although these metals are not absorbed into the animal system, these may excrete in the manure of animals (Guo et al., 2012). Wu et al. (2017) investigated the distribution of Zn and Cu in the composting of pig manures (PMs) and reported that the total contents of Zn and Cu were increased from 112.8% to 192.7% for Zn and from 115.5% to 132.6% for Cu at the end of composting process.

4.2 EFFECTS OF HEAVY METALS ON THE COMPOSTING PROCESS

Table 4.2 represents a setup of heavy-metal limits for compost to be applied in agriculture in different countries. The microbial diversity during the composting process may vary with the heavy metals present in the composting mixture. Microorganisms are strongly involved in the degradation of OM followed by detoxifying some

TABLE 4.2
Setup of Heavy Metal Limits for Compost to be Applied in Agriculture in Different Countries

	Concentration of Heavy Metals (mg/kg dry weight)							
Name of Countries	**Zn**	**Cu**	**Ni**	**Cd**	**Cr**	**Pb**	**As**	**Hg**
India	1000	300	50	5	50	100	10	0.15
Austria	1000	400	100	4	150	500	–	4
Belgium	1000	100	50	5	150	600	–	5
Switzerland	500	150	50	3	150	150	–	3
Denmark	–	–	45	1.2	–	120	25	1.2
France	–	–	200	8	–	800		8
Germany	400	100	50	105	100	150	–	1.0
Italy	500	300	50	105	100	140	10	1.5
Netherlands	900	300	50	2	200	200	25	2
Spain	4000	1750	400	40	750	1200	–	25
Canada	500	100	–	3	210	150	13	0.8

organic and inorganic pollutants, as well as changing the mobility and bioavailability of heavy metals to plants. Subsequently, microbial reproduction is affected by heavy metals, resulting in morphological and physiological changes. Therefore, the biodegradation of organic biomass could be influenced by toxic heavy metals in the composting process. Microbial enzymatic activities may be reduced by toxic heavy metals (Huang et al., 2006).

In the composting process, microbes and their secreted enzymes play a key role in the biological and biochemical conversions of the composting biomass (Guo et al., 2012). However, the antimicrobial action of copper may occur due to the ability of copper ions to chelate sulfhydryl groups, thus interfering with the cell proteins or enzymes (Ochoa-Herrera et al., 2011). The high concentration of Cu have negative effect on the microbial activities through their enzyme activity and consumption for carbon source in the composting mixture (Guo et al., 2012). The pollution-induced community tolerance (PICT) has been shown to be a versatile and illustrative method for the detection of Cu on the microbial community. The tolerance level can be measured through the observed elimination of sensitive microbial species. The species of microbes occurs in mesophilic phase can be replaced by more thermally tolerant species through their physiologically and genetically adaptation (Demoling and Bååth, 2008). Li et al. (2015) concluded that the C/N ratios and pH values were not affected by the amendment of Cu; however, the enzymatic activities, such as urease and hydrogenase, were significantly inhibited by the high-Cu exposure in the composting process. The high concentration of Cu may impose selective pressures on the microbial tolerance of Cu during the composting process. It has been considered that microbial tolerances may be transferred to the soil through manure application and this enhances the ecological risk of animal wastes polluted by the Cu. Heavy metals decrease the phosphatase synthesis during the composting process (Garcia et al., 1995). Microorganisms have to live with toxic Pb during their growth in the

Pb-contaminated mixture of various wastes and the exposure of the microorganism to metals inhibits microbial growth and activity (Huang et al., 2006).

Malley et al. (2006) studied the effect of heavy metals such as Cu and Zn on dehydrogenase and protease activity of the substrate during the vermicomposting process by adding three different dosages of Cu and Zn. It was also reported that the dehydrogenase activity was decreased in substrate by increasing Cu or Zn dosages. The dehydrogenase level was decreased with increasing heavy-metal dosage. In the composting process, heavy metals may deactivate enzyme activities by complexing the substrate, by reacting with protein-active groups of enzymes, or by reacting with the enzyme–substrate complex or indirectly by changing the microbial community which synthesizes enzymes (Singh and Kalamdhad, 2011). In the composting process, heavy metals in addition to binding with aromatic amino-acid residues in enzyme molecules can also cause oxidative damage of proteins by the induction of oxidative stress associated with the production of reactive oxygen species such as hydroxyl or superoxide radicals (Baldrian, 2003; Singh and Kalamdhad, 2011).

4.3 REDUCTION OF TOTAL HEAVY METALS AND THEIR BIOAVAILABILITY DURING THE COMPOSTING PROCESS

4.3.1 Physicochemical Methods

The thermophilic phase of the composting process, which is the first step of the composting process, may affect the exchangeable and carbonate fractions of the heavy metals. The oxic and anoxic conditions developed during the composting process may affect the reducible and oxidizable fractions of metals (Zorpas et al., 2000). According to He et al. (2009a,b), the exchangeable fraction of Pb increased during the later thermophilic stage, but dropped again in the cooling period. A percentage of residual fractions of Zn increased during the mesophilic and thermophilic phases but decreased in the cooling stage. The exchangeable and carbonate fractions of Cu only accounted for the small parts of total Cu. The concentrations of both these fractions increased, although a decrease tendency appeared during thermophilic phase of the composting process. Amir et al. (2005) reported that the highest proportions of metals were found to be associated with the residual fraction of total metals—a more stable form and considered not easily available for plant absorption. The amount of potentially bioavailable metals was found to be $<2\%$ of total metal. Table 4.2 shows the physicochemical and biological methods, reducing agent, and target metals.

Heavy metals containing compost quality can be improved by adding some specific chemicals or waste materials. Many studies have been carried out by using natural zeolite (Sprynskyy et al., 2007; Stylianou et al., 2008; Zorpas et al., 2008; Villasenor et al., 2011), lime/waste lime (Wong and Selvam, 2006; Singh and Kalamdhad, 2013a,b, 2014a), lime and sodium sulfide (Wang et al., 2008), and bamboo charcoal (BC; Chen et al., 2010b). These chemical compounds were used as amendments, which can improve compost quality by eliminating or changing mobile and easily available forms of metals. The easily bioavailable form of heavy metals causes more serious problems to the environment since they can be easily absorbed by plants and can enter the trophic chain or pollute the ground waters (Sprynskyy et al., 2007).

4.3.2 Reduction of Metals by Addition of Chemicals

4.3.2.1 Lime/Waste Lime and Sodium Sulfide Treatment

Lime/waste lime is generally considered to be the appropriate amendment materials for the SS composting process. It plays a significant role in decreasing the microbial content of SS (mainly pathogens) and the bioavailability of heavy metals. Consequently, it improves the composting process and biodegradability of the OM (Samaras et al., 2008). Immobilization of heavy metals was carried out by the addition of lime during the SS composting process (Wong and Selvam, 2006; Pardo et al., 2011). According to Fang and Wong (1999), a slower increase in temperature was observed in sludge composting with lime amendment as compared with that of the control which contains only SS. The small amount of lime added to the composting mixture provided a buffer against the decrease in pH during initial degradation and an appropriate quantity of Ca, which would improve the metabolic activity of microbes during composting (Fang and Wong, 1999).

Wong and Selvam (2006) reported that the addition of lime caused a significant reduction in water-soluble Cu, Mn, and Zn contents during the SS composting process. Fang and Wong (1999) studied composting of sewage sludge with different proportions of lime to reduce the bioavailability of heavy metals in the sludge compost and specified that the lime addition could be very effective for reducing heavy-metal bioavailability during the composting of SS due to lime will involve in forming little carbonate-soluble salts. Fang and Wong (1999) also reported that Ni showed different patterns of change with a decrease in Ni concentration in order to the increase in time for all treatments in comparison to other metals. Chiang et al. (2007) investigated the effect of different amendments, such as lime, coal fly ash, and natural zeolite, on heavy-metal bioavailability during the SS composting process. This study reported that coal fly ash resulted in a significant reduction in DTPA-extractable metals. Zeolite and lime addition generally decreased the DTPA-extractable heavy metals, such as Pb, Cu, and Zn, due to the higher ion exchange capacity and alkalinity in the composting process (Chiang et al., 2007) (Table 4.3).

Wong and Selvam (2006) described that the residual form of Zn was mainly transformed into its oxidizable form during the composting process and lime addition reduced this transformation. A similar trend was reported by Samaras et al. (2008), who showed that the oxidizable form of Zn was transformed into its residual form during the SS composting process with the addition of lime and fly ash, and lime addition reduced this transformation. Chemical forms (exchangeable, carbonate, reducible, oxidizable and residual) of metals generally influenced by the pH as well as OM.

Fang and Wong (1999) reported that the addition of lime caused a significant reduction in the water-solubility of Cu, Mn, and Zn in the SS composting process. Wang et al. (2008) reported that the addition of alkaline lime could neutralize the organic acids released during composting, thus reducing the formation of metal–OM complexes during lime-sludge co-composting. This study also reported that the addition of sodium sulfide and lime caused an increase in the oxidizable fraction of Cu during composting. Wong and Selvam (2006) reported that the exchangeable, carbonate, and reducible fractions were increased at the end of composting process, and the addition of lime reduced this transformation. Wang et al. (2008) reported that the residual fraction of Zn was found to be the dominant fraction in

TABLE 4.3
Physicochemical and Biological Methods, Reducing Agents, and Target Metals

Methods	Reducing Agents	Target Heavy Metals	Composting Materials	Authors
Physicochemical methods	Natural zeolite	Zn, Cu, Ni, Cr, Cd, Pb, Fe, and Mn	Sewage sludge	Zorpas et al. (2000)
		Zn, Cu, Ni, Cr, Cd, and Pb	Sewage sludge	Chiang et al. (2007)
		Cu, Ni, Cr, Cd, and Pb	Sewage sludge	Sprynskyy et al. (2007)
		Zn, Cu, Ni, Cr, Pb, and Mn	Sewage sludge	Stylianou et al. (2008)
		Cu, Ni, Cr, Cd, and Pb	Sewage sludge	Kosobucki et al. (2008)
		Zn, Cu, Ni, Cr, Cd, Pb, and Hg	Sewage sludge	Villasenor et al. (2011)
		Zn, Cu, Mn, Fe, Ni, Pb, Cd, and Cr	Water hyacinth	Singh et al. (2013)
		Zn, Cu, Mn, Fe, Ni, Pb, Cd, and Cr	Water hyacinth	Singh and Kalamdhad (2014b,c,d)
	Lime and sodium sulfide	Cu, Mn, Ni, and Zn	Sewage sludge	Fang and Wong (1999)
		Cu, Zn, and Ni	Sewage sludge	Wang et al. (2008)
		Zn, Cu, Ni, Cr, Cd, and Pb	Sewage sludge	Chiang et al. (2007)
	Waste lime	Zn, Cu, Mn, Fe, Ni, Pb, Cd, and Cr	Water hyacinth	Singh and Kalamdhad (2013b)
		Zn, Cu, Mn, Fe, Ni, Pb, Cd, and Cr	Water hyacinth	Singh and Kalamdhad (2014a)
		Zn, Cu, Mn, Fe, Ni, Pb, Cd, and Cr	Water hyacinth	Singh and Kalamdhad (2016)
	Bamboo charcoal and bamboo vinegar	Cu and Zn	Pig manure	Chen et al. (2010a)
		Cu and Zn	Sewage sludge	Hua et al. (2009)
	Red mud	Zn, Cu, Ni, Cr, and Pb	Sewage sludge	Qiao and Ho (1997)
	Coal fly ash	Zn, Cu, Ni, Cr, Cd, and Pb	Sewage sludge	Chiang et al. (2007)
Biological methods	Fungi	Pb	Lead-contaminated solid waste	Zeng et al. (2007)
	Bacteria	Pb, Cd, Cr, Mn, and Mg	Industrial sludge	Nair et al. (2008)
	Earthworm	Pb	Lead-contaminated solid waste	Liu et al. (2009)
		Pb, Cd, Cr, Cu, and Zn	Pig manure	Li et al. (2010)
		Zn, Cu, Cr, Cd, Pb, and Fe	Water hyacinth	Gupta et al. (2007)
		Zn, Cu, Mn, Fe, Ni, Pb, Cd, and Cr	Water hyacinth	Singh and Kalamdhad (2013c,d)
		Zn, Cu, Mn, Fe, Ni, Pb, Cd, and Cr	Phumdi biomass	Singh et al. (2014)
		Cr, Cu, Co, Zn, and Ni	Municipal solid waste	Soobhany et al. (2015)
		Zn, Pb, Cr, and Cu	Animal manure	Lv et al. (2016)

the lime treated compost. The exchangeable and carbonate fractions were decreased in sodium sulfide and lime amendments. It can be concluded that sodium sulfide and lime amendments would significantly decrease the mobility and bioavailability of metals during the composting process. Wang et al. (2008) also reported that the carbonate, reducible, and oxidizable forms of Ni were mainly transformed into the residual fraction during composting process. Singh and Kalamdhad (2013a,b) concluded that waste lime amendment resulted in a significant decrease in the water solubility of heavy metals (Zn, Cu, Fe, and Cr) and DTPA-extractable metals (Zn, Cu, Fe, Ni, and Cr) during the composting of water hyacinth. The highest reduction of metals were observed in lime treatment with 2% of total composting mass which represent the best percentage of lime, which could enhance OM degradation followed by formation of humus-like substances, resulting toxicity of the metals was reduced during water hyacinth composting.

4.3.2.2 Natural Zeolite Treatment

Natural zeolite has been used widely for reducing the bioavailability of heavy metals during the composting of SS, water hyacinth etc. (Zorpas et al., 2000; Sprynskyy et al., 2007; Singh and Kalamdhad, 2014b). Natural zeolite have ability to adsorbs and exchange with concerning the heavy metals. It has the ability to adsorb easily-available fractions of heavy metals and to exchange sodium and potassium with metal ions (Zorpas et al., 2000). Zeolites are naturally occurring hydrated aluminosilicate minerals and belong to the class of minerals known as "tectosilicates." The structure of zeolites consists of three-dimensional frameworks of SiO_4 and AlO_4 tetrahedra (Erdem et al., 2004; Villasenor et al., 2011). The net negative charge of zeolites is balanced by the exchangeable cations (Na, K, or Ca). These cations are exchangeable with certain cations in solutions such as Pb, Th, Cd, Zn, Mn, NH_4^+ etc. (Sprynskyy et al., 2007; Villasenor et al., 2011). Zorpas et al. (2000) reported that clinoptilolites have the ability to absorb the exchangeable and carbonate form of metals. Clinoptilolites adsorb exchangeable and carbonate fractions in the following sequences: Cu > Cr > Fe > Ni > Mn > Pb > Zn (Zorpas et al., 2000). Sprynskyy et al. (2007) reported that addition of the clinoptilolite rock to SS might change different fractions of heavy metals in composts and decrease their bioavailability during the composting of SS (Sprynskyy et al., 2007). Villasenor et al. (2011) stated that the zeolites take up 100% of Ni, Cr, and Pb in the composting process. This study also reported that Zeocat was the most effective form of zeolite for the removal of Cu, Zn, and Hg. Clinoptilolites had the ability to take up metals associated with mobile fractions (exchangeable and carbonate forms). Clinoptilolites allow metals in the following sequences: $Zn^{+2} > Cr^{+3} > Ni^{+2} > Cu^{+2} > Mn^{+2}$ (Stylianou et al., 2008). Zorpas et al. (2002) reported that the metals absorbed by clinoptilolites decrease with decreasing the particle size of clinoptilolites. Surface dusts that occur during the grinding process may hinder the metal uptake capacity of clinoptilolites by clogging the pores and causing physical impairments in smaller particles.

4.3.2.3 BC and Bamboo Vinegar (BV) Treatments

BC is composed of a large amount of micropores and has an enormously large surface area (approximately 10 times greater than those in the wood charcoal). BC has been considered a good candidate for the conservation of nutrient and immobilization of heavy metals due to its excellent adsorption capability for heavy metals (Hua et al., 2009). BC is produced by the transformation of bamboo under oxygen-limited conditions. BV is a brown–red transparent liquid produced during transformation of BC and contains approximately 200 types of chemical constituents, in which acetic acid is the main component of BV (Chen et al., 2010b).

Chen et al. (2010b) studied the effects of BC or BC + BV on temperature, pH, and other parameters of the composting of PM. This study also reported that the addition of BC + BV shortened the time of mesophilic stage, quickly developed the thermophilic phase, and decreased the pH value at the thermophilic phase.

Chen et al. (2010b) stated that approximately 74% of total nitrogen loss and 9% of bioavailability of Cu and Zn decreased with increasing BC + BV and BC, respectively, in the composting of PM. Hua et al. (2009) reported that mobility and bioavailability of heavy metals decreased with an increase of BC in the SS composting. However, the effect of BC on immobilization was not similar for Cu^{2+} and Zn^{2+}. DTPA-extractable Cu and Zn contents of the composted sludge decreased by 27.5% with BC amendment and 8.2% without BC amendment, respectively.

4.3.2.4 Red Mud Treatment

Red mud is an industrial waste by-product that is generated during the refining of bauxite to alumina through the Bayer's process (Gadepalle et al., 2007). Red mud is used effectively for the immobilization of heavy metals because it adsorbs metal cations from the aqueous solution due to its high pH and cation exchange capacity (Qiao and Ho, 1997).

The red mud is composed primarily of the following components: hematite, boehmite, quartz, and sodalite (and gypsum) (Gadepalle et al., 2007). The addition of red mud in the SS composting process generally reduces metal bioavailability and leachability and therefore the potential hazards of releasing metals thorough adsorption and complexation of the metals on to the inorganic components of red mud (Qiao and Ho, 1997). Red mud may also increase the fractionation of the pH, solid-to-solution ratio, and available adsorption sites etc. (Qiao and Ho, 1997).

Qiao and Ho (1997) reported that red mud reduces the bioavailability and leachability of heavy metals in the SS composting process. This study also reported that the oxidizable fraction of Cu was mainly transformed into exchangeable, carbonate, and reducible fractions. The addition of red mud in the composting mixture decreased the exchangeable fraction and increased the pH to precipitate Cu, whereas addition of inorganic oxide surface to adsorb Cu decreased the bioavailability of metals during SS composting. This study also reported that addition of red mud significantly reduced the leachability and plant availability of metals during SS composting (Qiao and Ho, 1997).

4.3.3 Biological Method

Development of cost-effective substitutes, such as biosorption, bioaccumulation, and bioleaching, has become a demanding area of manipulation over the past decade. Microbial biomass has been used due to its sorption capabilities (Siloniz et al., 2002).

Biosorption can be defined as the ability of a nonliving organism to accumulate heavy metals from waste water through a non-metabolically energetic process, whereas the bioaccumulation can be defined as an active process where the removal of metals involves the metabolic activity of a living organism (Philippis and Micheletti, 2009). Bioleaching is an efficient and economical process for the removal of heavy metals from the soils, sediments, industrial wastes, and solid wastes (Pathak et al., 2009). Bioleaching is an eco-friendly technique and is 80% cheaper than traditional chemical methods—engaged for the metals leaching from the sludge and recovery of metals from the leachate of different wastes—in terms of chemical feeding (Pathak et al., 2009).

Mechanical treatment of waste by grinding, mixing, and sieving out nondegradable materials (metals, plastics, glass, stones) provides good conditions for biological treatment of compostable materials (Barker et al., 2002).

According to Ahmad et al. (2005), microorganisms such as bacteria, fungi, algae, and yeast have been proven to tolerate and accumulate heavy metals. Microorganisms (e.g., bacteria, fungi, algae, and yeast) have been used for the treatment of wastewater contaminated with metal and could be expected to restrain metals in the composting of solid wastes (Zeng et al., 2007). The following mechanisms are involved in bio-adsorption process: ion exchange, coordination, complexation, chelation, adsorption, micro-precipitation, diffusion via cell walls, and membrane. This mechanism may vary with different species of microorganisms and depends on the species used, their origin, processing of the biomass, and solution chemistry (Ahmad et al., 2005). Shukla et al. (2009) stated that microbial communities play an important role in the detoxification, stabilization, and transformation of Cr in the composting biomass to ensure the environmental sustainability.

4.3.3.1 Application of Fungi

Zeng et al. (2007) reported that white-rot fungi can accumulate heavy-metal ions in their intra/inter cells and can be responsible for metal chelation through the carboxyl, hydroxyl, or other active functional groups on the surface of cell walls (Zeng et al., 2007). Moreover, white-rot fungi can accumulate metals from substrate using their mycelia (Baldrian, 2003). Huang et al. (2006) reported that *Phanerochaete chrysosporium* is a good candidate for absorbing metal ions from the diluted solutions through its mycelium and it has potential to develop in both the solid and liquid environments. Thus, it may reduce various xenobiotic compounds not only in nutrient-rich environments but also in the nutrient-limited condition. Extracellular enzymes of fungi can directly interact with metals present in the environment (Baldrian, 2003). Liu et al. (2009) reported that the composting methods with inoculants of *P. chrysosporium* could efficiently transform Pb fractions and reduce active Pb concentration in composting biomass. The transformation behavior of Pb fractions might result from the fact that the Pb ions could be accumulated by fungal mycelium and chelated by the humus formed in the composting. The exchangeable fraction of Pb is strongly correlated with pH and microbial biomass, and increasing the pH of composting biomass and microbial biomass is an important factor for the stabilization of Pb during the composting process.

4.3.3.2 Using Bacteria

The thermophilic bacteria found in solid waste compost belong mainly to the genus *Bacillus B. stearothermophilus*, which is also known as the major dominant species of the thermophilic phase (at temperature >65°C) of the composting process, while *Thermus* strains plays an important role in OM degradation during the thermogenic phase (at temperature >70°C) of the composting process (Fang and Wong, 2000). Siderophores are defined as low molecular weight ligands synthesized and excreted by bacteria for capturing and transporting iron to support metabolic activity (Nair et al., 2008). The metal-sulfide-dissolving microorganisms are enormously acidophilic bacteria (flourishing at pH <3). These microorganisms can oxidize either inorganic sulfur compounds and/or Fe (II) ions. The conventional leaching bacteria belong to the genus *Acidithiobacillus* (formerly *Thiobacillus*) (Rohwerder et al., 2003).

4.3.3.3 Using Earthworms

Vermicomposting is a low-cost technology system that primarily uses earthworms for the degradation of organic wastes. It is a well-known method managed by earthworms which can reduce toxicity of heavy metals through accumulation in their body parts and formation of organometallic complex during the process (Ghyasvand et al., 2008). Vermicomposting using earthworm accelerates OM stabilization and excrete chelating and phytohormonal substances which have a high content of microbial matter and stabilized HS (Gupta and Garg, 2008; Suthar, 2009, Hait and Tare, 2012). The vermicomposting process involves the stabilization of organic material by the joint action of earthworms and microorganisms. Earthworms are considered to be the main factors for providing the favorable condition of the substrate and shifting the biological activity, whereas microbes are accountable for biochemical degradation of OM (Vig et al., 2011).

Li et al. (2010) reported that *Eisenia fetida* can accumulate Cu, Zn, Pb, and Cd. The adult earthworms have an ability to store high concentrations of heavy metals in the nontoxic forms. *Eisenia fetida* can accumulate soluble and exchangeable metals easily (Li et al., 2010). Earthworms tend to increase the solubilization of Fe and this behavior of earthworms can lead to the degradation of OM in the presence of high content of different microorganisms inside the gut of an earthworm (Bhattacharya and Chattopadhyay, 2006). Liu et al. (2009) reported that the soluble and exchangeable Pb contents decreased at higher pH values. The pH is known to affect the ionic form and chemical mobility, so a high pH value can decrease the solubility of metal ions in the medium.

Hait and Tare (2012) reported that *E. fetida* can decrease water solubility of Cu, Zn, and Cr. The mechanism for reducing metal contents during the vermicomposting process is as follows: when the OM is passed through the earthworm's gut, some part of it is digested, as pH and microbial activity in the gut was increased. Earthworm cast have binding ability of metals ions with negative ions which are present in it.

Hait and Tare (2012) reported that that water solubility of Cr decreased during the vermicomposting process is possibly due to the converting Cr(VI) to Cr(III) by earthworm as well as microorganisms, consequently, Cr(III) is combined with decomposed organic materials. There are two types of mechanisms which may be involved in the reduction of mobility and bioavailability of heavy metals: the first one is the binding of metals to nuclear proteins and the formation of inclusion nuclear bodies and the second one is a cytoplasmic process involving the formation of a specific metal binding protein and metallothionein within the chloragogenous tissue (Hait and Tare, 2012). The interaction of the humic acid with metals is one of the main factors affecting the fractionation of heavy metals. Humic acid has a stronger sorption effect on Cu and Zn (Kang et al., 2011). It has been also suggested that the bioavailability of heavy metals tends to decrease during the vermicomposting process due to the formation of an organometallic complex (Dominguez and Edwards, 2004). Additionally, the epithelial layer of the gut absorbs the necessary contents of toxic metals when it passes through the gut and can accumulate metals in earthworm tissues (Suthar et al., 2008; Suthar and Singh, 2009). Jain et al. (2004) reported that the heavy-metal contents were reduced during the vermicomposting process and in *E. fetida* the mitochondrial and cytoplasmic fractions can convert a highly toxic form of Cr (VI) to a nontoxic form of Cr (III). Suthar and Singh (2008) stated that metal reduction during the vermicomposting process was related to the earthworm activity in the wastes used for vermicomposting. This study also reported that earthworms have a potential to accumulate heavy metals in their tissues. Suthar (2009) reported that DTPA-extractable metals, such as Cu, Fe, Mn, and Zn, were reduced during the vermicomposting process of SS spiked with sugarcane trash. The bioaccumulation of the water-soluble fraction of metals can be increased when it passes through the guts of earthworms (Suthar, 2009). Rorat et al. (2017) reported that Cr was not found in control worms, whereas it occurred in all worms used in the experiment. Concentrations of Cu and Ni were generally found to be higher in experimental worms than those in control worms, whereas no differences were found in specific experiments. Body accumulation factor exceeded 1 only for Cd (17.44). Bioaccumulation factors (BAFs) calculated for all analyzed metals can be ranked in the following sequences: Cd > Cu > Zn > Ni > Cr > Pb. Thus, BAFs can be calculated as follows (Rorat et al., 2017):

$$BAF = \frac{\text{CM earthworm}}{\text{CM soil}}, \tag{4.1}$$

where CM earthworm is the total concentration of a selected metal in an earthworm's body (mg/g) and CM soil is the total concentration of the same metal in substratum (mg/g).

Lv et al. (2016) concluded that vermicomposting reduced the mobility of Zn, Pb, and Cu through changing their chemical speciation into less-available forms; however, their total concentration of Zn, Pb, and Cu were increased. The speciation of heavy metals in the vermicomposting process were changed by earthworm's activities. The bioaccumulation of metals by earthworms and the change of substrates properties, resulting transformation and reduction in mobility of heavy metals happened during the vermicomposting process.

There are two types of fractionation procedures: (i) water-soluble fraction and (ii) DTPA-extractable fraction: potential mobilizable. DTPA can be used as a chelating agent containing an aminopolycarboxylic acid with diethylene triamine backbone and five carboxymethyl groups having the capability to extract exchangeable, carbonate, and organically bound metal fractions, such as potentially mobile and phytotoxic fractions (Garcia et al., 1995; Fang and Wong, 1999; Chiang et al., 2007).

A significant reduction in the DTPA-extractable fractions has been reported during vermicomposting of various wastes (Suthar, 2008, 2009; Singh and Kalamdhad, 2013c). For example, Suthar (2008) reported a significant reduction in DTPA-extractable metals in the following sequences: 12.5%–38.8% for Zn, 5.9%–30.4% for Fe, 4.7%–38.2% for Mn, and 4.5%–42.1% for Cu during the vermicomposting of distillery sludge. Suthar (2009) reported a significant reduction in the metals in the following sequences: 4.5%–30.5% for Cu, 5.1%–12.6% for Fe, 3.3%–18.0% for Mn, 2.5%–15.9% for Zn, and 2.4%–20.0% for Pb during vermicomposting of SS mixed with sugarcane trash. Singh and Kalamdhad (2013c) reported a significant reduction in the metals in the following sequences 4.0%–33.6% for Zn, 9.2%–57.7%for Cu, 36.5%–54.6% for Fe, 63.7%–73.7% for Ni, and 15.2%–51.3% for Cr in vermicomposting of water hyacinth mixed with cattle manure. Singh and Kalamdhad (2013d) concluded that the bioavailability of heavy metals depends mainly on the physicochemical and biological properties of the vermicomposting process. The pH value was increased significantly with a decrease in the bioavailability of the heavy metals at the end of vermicomposting process. The exchangeable fraction of Mn was found to be dominant in the initial feed mixture (water hyacinth mixed with cattle manure and sawdust), whereas in the final vermicomposting process it was converted into less-mobile fractions such as reducible, oxidizable, and residual fractions. The exchangeable fraction of Cd was decreased approximately 100% during the vermicomposting process. The total concentration of Pb was higher than those of Zn, Cu, Mn, Ni, Cd, and Cr; however, its bioavailability factor was very low in the all eight metals. Ni, Cd, Pb, and Cr were found mainly in the residual fraction in the final vermicomposting process.

Eisenia fetida have potential to accumulate exchangeable and carbonate-bound fractions of Zn (Li et al., 2010). The exchangeable and carbonate fractions of metals were decreased during the vermicomposting process possibly due to the decomposition and conversion of organic materials into insoluble salts or HS to form a stable form of OM in the composting/vermicomposting (Song et al., 2014; Lv et al., 2016). The transformation of heavy metals during the vermicomposting process can be explained as microbial community inside gut of worms encouraged by earthworms which could degrade the OM enzymatically, resulting release the metals in the vermicomposting process, and then released metals are combined with the humus (Sizmur and Hodson 2009; Wang et al., 2013). The substantial formation of HS might be responsible for increasing the organically bound fraction (Hsu and Lo, 2001; Tandy et al., 2009). Goswami et al. (2014) reported that heavy metals can prompt the production of metallothionein protein in gut of earthworm, then it bind metal ions by forming organometallic ligands, resulting and thus reduce the exchangeable fraction of metals. According to Yang et al. (2013), the extractable heavy metals are immobilized by the formation of insoluble salt-like phosphates. Wang et al. (2013) reported that a significant reduction in the bioavailability of metals in the following sequences: 6.0%–8.7% for Cu, 6.8%–10.3% for Zn, 5.6%–12.7%

for Pb, 4.3%–13.6% for Cd, and 1.0%–2.1% during the vermicomposting of SS with fly ash and phosphate rock.

Singh and Kalamdhad (2013d) studied speciation of heavy metals during the vermicomposting of water hyacinth. This study also reported that exchangeable, carbonate, and reducible fractions of Zn were found to decrease in the range of 32.5%–86%, 13.3%–27.8%, and 2.9%–71.3% of the total fractions, respectively, after the vermicomposting process. However, fractions associated with OM and residual fractions of Zn were increased after the vermicomposting process. The bioavailability factor of Zn was found to decrease from 0.62 (initial) to 0.59 (final) during the vermicomposting process. Lv et al. (2016) stated that Zn is highly sensitive to the change in pH. The carbonate fraction of Zn was decreased significantly, whereas its reducible and oxidizable fractions were increased in both cattle manure and PM vermicomposting. The residual fraction of Zn had similar effects with the exchangeable fraction in both cattle manure and PM vermicomposting. In addition, in the residual fraction no significant differences were observed between vermicomposting and the control (without the addition of earthworms). Furthermore, exchangeable and carbonate fractions in vermicomposting of both cattle dung and PM were significantly decreased as compared with the control. Lv et al. (2016) studied chemical speciation of heavy metals during vermicomposting of cattle manure (cattle dung and PM). This study also reported that the exchangeable fraction of Zn was significantly decreased 76.67% during cattle dung vermicomposting, whereas it was increased approximately 38.6% in PM vermicomposting due to the decrease of pH after PM vermicomposting process. He et al. (2016) studied speciation of heavy metals in the vermicomposting of SS with additive materials. This study also reported that the exchangeable fraction of Zn was decreased approximately 38.5% of the total fraction, whereas the exchangeable and residual fractions of Zn were increased and the reducible and oxidizable fractions were decreased. Wang et al. (2013) reported that the exchangeable fraction of Zn was decreased through the vermicomposting of each treatment, whereas the concentrations of Zn in the other fractions (e.g., carbonate, reducible, oxidizable, and residual fractions) in raw material mixtures were found to be somewhat changed after vermicomposting of SS.

Lv et al. (2016) reported that the exchangeable and carbonate fractions of Pb were significantly decreased in *E. fetida*, whereas the exchangeable and carbonate fractions of Pb were significantly decreased in both cattle dung and PM vermicomposting as compared with the control. However, the reducible and organically bound fractions of Pb were evidently increased during vermicomposting. Singh and Kalamdhad (2013d) stated that exchangeable fraction of Pb was reduced approximately 15.7% of the total fraction and it contributed in the range of 0.8%–1.5% of the total fraction. The carbonate fraction of Pb was increased, whereas the reducible and organically bound fractions of Pb were not found in the final vermicompost. The residual fraction of Pb was dominant from initial to final vermicompost and it was decreased after vermicomposting. Bioavailability factor of Pb was increased during the vermicomposting process. Li et al. (2010) found approximately 15% exchangeable fraction of the total concentration of Pb in the final vermicompost of PM. He et al. (2016) reported that the exchangeable fraction of Pb was decreased, whereas its residual fraction was increased. Wang et al. (2013) stated that the highest proportion of Pb was found to be associated with the oxidizable and residual fractions of total Pb before and after the vermicomposting process.

Singh and Kalamdhad (2013d) stated that exchangeable, carbonate, and reducible fractions of Cr were decreased approximately 66%, 72%, and 82% of total fractions, respectively, at the end of the vermicomposting process, whereas the oxidizable and residual fractions of Cr were increased approximately 23% and 25% of total fractions. Cr was mainly distributed in oxidizable and residual fractions. The bioavailability factor of Cr was decreased from 0.37 (initial) to 0.27 (final) during the vermicomposting process. This decrease in the bioavailability factor of Cr might be due to bioaccumulation of the free form of Cr by *E. fetida* (Singh and Kalamdhad, 2013d). Lv et al. (2016) reported that most of the Cr fractions identified belonged to the organic and residual fractions in cattle dung and PM. The exchangeable and carbonate fractions of Cr were decreased approximately 35.3% and 22.2% during the vermicomposting process of both cattle dung and PM, whereas these fractions were increased in the control (without the addition of earthworms). Hence, this study demonstrates that vermicomposting can reduce the eco-toxicity of Cr during vermicomposting. The reducible fraction of the Cr increased suggestively after the vermicomposting process. The addition of earthworms decreased the residual fraction of Cr in the cattle dung vermicomposting as compared with the control, whereas its residual fraction increased in PM vermicomposting (Lv et al., 2016). He et al. (2016) reported that the exchangeable fraction of Cr was decreased, whereas its residual fraction was increased approximately 68%–88% after the vermicomposting.

Singh and Kalamdhad (2013d) reported that the exchangeable fraction of Cd was decreased approximately 100% of the total fraction during the vermicomposting process at the higher pH values 6.8 (initial) to 7.7 (final), whereas the carbonate fraction of Cd was increased. The reducible and organically bound fractions of Cd were not found during the vermicomposting process. The residual fraction of Cd was dominant from initial to final vermicompost, whereas its residual fraction was increased approximately 2.5% in the final vermicompost. He et al. (2016) reported that the exchangeable and residual fractions of Cd were increased, whereas the reducible and oxidizable fractions of Cd were decreased significantly. Wang et al. (2013) stated that the exchangeable fraction of Cd was considered to be the dominant fraction of the total Cd during the vermicomposting process.

Li et al. (2010) reported that Cu was absorbed/accumulated by *E. fetida* and accumulation of Cu have a strong positive correlation with the concentration of exchangeable fraction. Lv et al. (2016) reported that the exchangeable fraction of Cu was significantly decreased in both cattle dung and PM during the vermicomposting process, whereas its organically bound fraction was significantly increased in both cattle dung and PM during the vermicomposting process. Singh and Kalamdhad (2013d) reported that exchangeable and carbonate fractions of Cu were decreased in the range of 11.4%–67% and 20.6%–48.8% of total fractions, respectively. The oxidizable and residual fractions of Cu were increased after vermicomposting. The reduction in exchangeable and carbonate fractions might be due to the accumulation of these fractions by earthworms and transformation into other stable fractions (oxidizable and residual fractions). The oxidizable and residual fractions of Cu were found to be dominant in the final vermicomposting of water hyacinth. The bioavailability fraction of Cu was decreased from 0.59% (initial) to 0.55% (final) during the vermicomposting process. Wang et al. (2013) reported that the highest proportion of Cu was found to be associated with the oxidizable and residual fractions of total Cu in the vermicomposting mixtures, whereas the exchangeable fraction of Cu was found to be associated with a small proportion of the total Cu. The exchangeable fraction of Cu was decreased after vermicomposting. He et al. (2016) reported that the residual fraction of Cu was approximately 6.9% before the vermicomposting process, whereas this fraction was increased up to 38.1% and the exchangeable fraction of Cu did not show any change. The oxidizable and reducible fractions of Cu were decreased significantly during the vermicomposting. The increase in the residual fraction might be due to the transformation of oxidizable and reducible fractions into the residual fraction (He et al., 2016).

Singh and Kalamdhad (2013d) reported that the exchangeable fraction of Fe was increased up to 140% of the total fraction and it contributed in the range of 0.2%–0.9% of the total fraction of Fe in the water hyacinth vermicomposting. The carbonate fraction of Fe was decreased at the end of the vermicomposting. The reducible and oxidizable fractions of Fe were increased approximately 39% and 40.8% of the total fraction, respectively, whereas the residual fraction of Fe was decreased 63.2% during the vermicomposting process. The oxidizable fraction of Fe was considered dominant in the vermicomposting of hyacinth. The bioavailability factor of Fe was possibly increased due to its poor organometallic complex formation during the vermicomposting. He et al. (2016) reported that the exchangeable fraction of Fe was decreased, whereas its residual form was increased.

Singh and Kalamdhad (2013d) reported that the exchangeable and carbonate fractions of Mn were decreased in the range of 18%–51.4% and 0.6%–11.3% of total fraction, respectively. The exchangeable fraction of Mn was considered dominant in the final vermicomposting. This study also reported that the reducible and oxidizable fractions of Mn were increased in the final vermicomposting process, whereas its residual fraction was decreased during the vermicomposting process. Moreover, the bioavailability fraction of Mn was increased in the final vermicomposting of water hyacinth. He et al. (2016) reported that the exchangeable and residual fractions of Mn increased, whereas the reducible and oxidizable fractions of Mn decreased.

Singh and Kalamdhad (2013d) stated that the exchangeable and carbonate fractions of Ni were decreased approximately 56% and 48% of the total fraction. These fractions contributed <5% of the total fraction of Ni. The reducible and oxidizable fractions of Ni were not observed but the residual fraction was observed and increased after the vermicomposting. The residual fraction of Ni was found to be dominant throughout the vermicomposting process. This study also reported that the exchangeable and carbonate fractions of Ni were transformed into the residual fraction during the vermicomposting. The bioavailability factor of Ni was found to be in the range of 0.048%–0.027% during the vermicomposting process. He et al. (2016) reported that the residual fraction of Ni was found to be in the range of 19.7%–22.6% in raw materials, whereas this fraction was two times higher in the final vermicompost. The exchangeable fraction of Ni was decreased approximately 44%–26%

and only minor changes were seen in the reducible and oxidizable fractions of Ni during the vermicomposting process. From this study, it can be concluded that the water-soluble fraction of Ni decreased significantly in the vermicomposting process.

He et al. (2016) reported that the residual fraction of As was found to be approximately 53.3% in the initial stage of vermicomposting process, whereas its residual fraction was increased to 70.2% after the vermicomposting process. This study also reported that the speciation behavior of As changes irregularly among various treatments of vermicomposting of sludge. Wang et al. (2013) reported that the bioavailability factor of As significantly decreased in the vermicomposting of SS with phosphate rock, whereas the bioavailability factor of As significantly increased after the addition of fly ash or a mixture of fly ash with phosphate rock in the control. The bioavailability factor of heavy metal might be decreased due to an interaction between the HS and heavy metals after the vermicomposting process. The exchangeable and residual fractions of As were found to be dominant before and after the vermicomposting process.

4.4 RISK ASSESSMENT OF HEAVY METALS DURING THE COMPOSTING

The quantitative characterization and environmental quality of final compost can be assessed by using the following three criteria: risk assessment code (RAC), reduced partition index (IR), and potential ecological risk index (RI). Kulikowska and Gusiatin (2015) reported that the bioavailability factor of Zn was increased significantly (approximately 43.3% [series 1] and 39.8% [series 2]) in the SS composting process. This study also reported that the highest proportion of Zn was found to be associated with the exchangeable and acid-soluble fractions of total Zn in the SS composting process. However, the mobility factor of Zn was found to be <30% due to its transformation into more stable fractions during the composting process. The mobility factor of Cu was found to be <13%, whereas the mobility factors of Ni, Pb, and Cr were found to be <1.5%. Thus, it can be concluded that the mobility fractions of these metals were decreased due to their transformation into more stable chemical fractions in the composting process.

4.4.1 RAC

The RAC can be a useful means to evaluate the environmental risk associated with heavy-metal pollution in the final compost of various wastes. This method has been widely used by several researchers for assessing heavy-metal pollution in compost/sediments/soils (Sundaray et al., 2011). RAC is used to measure the bioavailability of metals as exchangeable and carbonate fractions, because these fractions are not strongly bonded, easily influenced by ionic strength, and susceptible to pH changes in the soil/compost (Yuan et al., 2011). The exchangeable and carbonate fractions may be dissolved in the aqueous phase due to the high mobility of these fractions and consequently become more quickly bioavailable (Sundaray et al., 2011). Table 4.4

TABLE 4.4
Evaluation of Risk Assessment Code (RAC) in Compost/Soil/Sediment

Category	Risk	Value of RAC
I	No risk	<1
II	Low risk	1–10
III	Medium risk	11–30
IV	High risk	31–50
V	Very high risk	>50

classifies the risk characterized in terms of RAC value. RAC is calculated as follows (Sundaray et al., 2011):

$$\text{RAC} = \left(\frac{\text{Exchangeable} + \text{Carbonate}}{\text{Total content of heavy metals}} \right) \times 100 \quad (4.2)$$

Singh et al. (2015) reported that all the metals showed low risk in the final compost of water hyacinth. The heavy-metal risk was greatly decreased in the final compost of lime treatments 1 and 2 (1% and 2% treatment), particularly for Fe, Ni, Pb, Cd, and Cr, with their risk decreased from low to zero. The risk of Zn continued at a low level, although the RAC values suddenly declined from 11.8% to 5.0% in control, 11.4% to 1.4% in lime treatment of 1%, 17.1% to 4.0% in lime treatment of 2%, and 12.3% to 4.9% in lime treatment of 3%. The risk of Cu was also observed at a low level. The decrease of RAC shows that the easily bioavailable fraction of heavy metals such as exchangeable and carbonate were mainly converted into the comparatively stable heavy-metal fractions such as reducible, oxidizable, and residual at the end of the composting process of water hyacinth, resulting in a decrease in the toxicity and environmental risk of the heavy metals (Singh et al., 2015). Yan et al. (2018) reported that the average RCA values of Cu, As, Pb, and Cr were found to be in the range of 1.9%–4.3% in each of the raw substrates or digestates with no risk or low risk, whereas the average RAC values of Zn, Ni, and Cd were found be in the range of 23.0%–55.8% with moderate-to-very high risk. Cd had relatively lower RAC values with no risk or low risk. Mn had average high RAC values of 89.2% with very high risk.

Gusiatin and Kulikowska (2014) reported that Cu, Ni, and Pb had lower RAC values with low risk, whereas the RAC values for Zn in all the composts were found to be significantly higher in the range 24.4%–39.1% with medium and high risk. A very high concentration of metals are present in their bioavailable form with low concentration of total metal in compost should not pose a serious apprehension in contrast to low concentration of bioavailable metal in a highly polluted waste/compost (Gusiatin and Kulikowska, 2014).

4.4.2 IR

The IR is a widely used parameter designed to measure the metal stability in composts. The IR uses the results of sequential extraction to describe the relative binding intensity of metals and is defined as follows (Gusiatin and Kulikowska, 2014):

$$I_R = \frac{\sum_{n=1}^{k} \left(i^2 F_i\right)}{k^2}, \tag{4.3}$$

where i is the index number of the extraction step, progressing from 1 (for the weakest) to the strongest fraction, k is extraction step, and F_i is the percentage of a particular metal present in fraction i.

The IR was introduced to quantitatively describe the relative binding intensity of metals in compost/soil and allows the comparison of the binding intensities of a selected metal in unlike soils/compost or of different metals in the same soil/compost (Miretzky et al., 2011; Gusiatin and Kulikowska, 2014). This index can be applied for each sequential extraction procedure. Zhu et al. (2012) proposed using a modified potential ecological risk index (MRI) that multiplies the RI index by a toxic index resulting from the RAC. The MRI has been known as useful for the risk management of heavy metals in sediments/compost, because it comprises total metal content, fractionation of metal, and toxic effect of most mobile fractions of metal. Gusiatin and Kulikowska (2014) reported that the relative binding intensity of Cu, Zn, Ni, and Pb changed previously at the beginning of composting, which is closely related to the distribution pattern of the metals. These changes in IR values for individual metals were caused by their redistribution among labile and stable fractions during the SS composting.

The IR value close to 0 represents very high mobility of metals, whereas its value close to 1—shows very high stability of metals (Han et al., 2003). Kulikowska and Gusiatin (2015) reported that reduced partition index (IR) reflecting on Cu, Ni, Pb, and Cr in mature composts ($0.65 < \text{IR} > 0.89$) represents the stability of metals. This study also reported that due to very low concentration of Cd and Hg in the final compost, their mobility and stability are impossible measured (on the basis of sequential extraction). The RI value was found to be <16 due to low concentrations of heavy metals such as Cd, Cr, Cu, Hg, Ni, Pb, and Zn, which represents the suitability and sanitary quality of the compost and can be applied as a soil amendment.

4.4.3 Potential Eecological RI

RI can be evaluated on the basis of the chemical speciation distribution of heavy metals (Zhai et al., 2014). It has been widely used for assessing the heavy-metal pollution and can be calculated as follows (Devi and Saroha, 2014; Yan et al., 2018).

$$C_f^i = \frac{C_i}{C_n^i}, \tag{4.4}$$

$$E_r^i = T_r^i \times C_f^i, \tag{4.5}$$

$$\mathrm{RI} = \sum_i^n E_r^i, \tag{4.6}$$

where C_f^i and C_i are the contamination factor and measured content for each heavy metal, C_n^i is the background values of specific heavy metal defined by Chen et al. (1991), E_r^i is the monomial potential ecological risk factor for specific heavy metal, and T_r^i is the toxic factor of individual heavy metal. The values of T_r^i used for assessing RI for individual heavy metals are as follows: 1 for Mn and Zn separately, 2 for Cr, 5 for Cu, Pb, and Ni separately, 10 for As, and 30 for Cd (Yan et al., 2018). According to Maanan et al. (2015), $E_r^i < 40$ denotes a low risk, $40 < E_r^i < 80$ a moderate risk, $80 < E_r^i < 160$ a considerable risk, $160 < E_r^i < 320$ a high risk, and $E_r^i > 320$ a very high risk. According to Fu et al. (2014), RI < 150 represents low risk, 150 < RI < 300 moderate risk, 300 < RI < 600 considerable risk, and RI > 600 very high risk.

4.5 CONCLUSION

Organic waste composting is one of the most economically feasible techniques to convert organic material into a stable, humic-like substance that can be used as a soil amendment. The concentration of heavy metals increases during the composting process. Heavy metals present in the composting mass can affect microbial activity within the composting biomass. The composting followed by land application is one of the most inexpensive methods for treatment and final disposal of SS, industrial sludge, water hyacinth, tannery sludge due to recycling of nutrients, and management of biomass disposal. Heavy metal absorption by plants tends to accumulate in the living tissues through the food chain causing problems to both human health and the environment. Heavy metals may enter into aquatic environment through agricultural runoff, and subsequently harm to aquatic environment. Therefore, composts containing metals and pathogens should not be used in agricultural soils. The composting process can decrease the toxicity of metals in terms their water solubility, DTPA extractability, and leachability. The composting process can encourage the complexation of heavy metals with OM resulting in a decrease in the mobility and availability of metals. pH is also an important parameter which may affects mobility and bioavailability of metals in the composting process. Analysis of heavy metals in the composts is very important for the routine monitoring, risk assessment, and regulation of environment. The addition of some chemicals, microbial inoculants, and earthworms to the composting process decreased the metal concentration in the composting process. Addition of lime in the composting biomass is enhanced degradation of OM and maintained neutral pH of final compost. Lime provide calcium to the composting microorganisms, resulting the metabolic activities of microorganisms are increased, subsequently reduced bioavailability of heavy metals. Natural zeolites have the potential to immobilize metals due to their high ion exchange abilities for Na and K with metals and highly porous structures. BC has a negative surface charge and a high charge density. These specific properties help BC in reducing bioavailability of heavy metals during the composting process. Red mud can change fractionation of metals through raising the pH value, solid-to-liquid ratio, and through availability of adsorption sites. The inoculation of microorganisms is also very important with respect to the reduction of metal contents by enhancing enzymatic activities, and the resulting quality of the compost is acceptable with low concentrations of heavy metal. Earthworms can accumulate a high level of heavy metals in less/nontoxic forms and also have potential to reduce probable toxic effects of surplus heavy metals through the utilization of metals for their physiological metabolism. The exchangeable fraction of Cd was decreased 100% during the vermicomposting of water hyacinth using *E. fetida*. The sequence of the bioavailability factors of different metals in final vermicomposting of water hyacinth is as follows: Mn (0.9) > Fe (0.86) > Zn (0.7) > Cu (0.6) > Cr (0.5) > Ni (0.07) > Cd (0.03) > Pb (0.02). Vermicomposting of metal containing wastes by using *E. fetida* was highly effective for the reduction of bioavailability of heavy metals. The biological process is the low-cost process as compared with the chemical method.

REFERENCES

Ahmad, I., Zafar, S., and Ahmad, F. 2005. Heavy metal biosorption potential of Aspergillus and Rhizopus sp. isolated from wastewater treated soil. *Journal of Applied Science and Environmental Management* 9(1): 123–126.

Amir, S., Hafidi, M., Merlina, G., and Revel, J.C. 2005. Sequential extraction of heavy metals during composting of sewage sludge. *Chemosphere* 59: 801–810.

Baldrian, P. 2003. Interactions of heavy metals with white-rot fungi. *Enzyme and Microbial Technology* 32: 78–91.

Barker, A.V., and Bryson, G.M. 2002. Bioremediation of heavy metals and organic toxicants by composting. *Scientific World Journal* 2: 407–420.

Bhattacharya, S.S., and Chattopadhyay, G.N. 2006. Effect of vermicomposting on the transformation of some trace elements in fly ash. *Nutrient Cycling in Agroecosystems* 5: 223–231.

Chen, J., Wei, F., Zheng, C., Wu, Y., and Adriano, D.C. 1991. Background concentrations of elements in soils of China. *Water Air and Soil Pollution* 57–58(1): 699–712.

Chen, G.Q., Chen, Y., Zeng, G.M., Zhang, J.C., Chen, Y.N., Wang, L., and Zhang, W.J. 2010a. Speciation of cadmium and changes in bacterial communities in red soil following application of cadmium-polluted compost. *Environmental Engineering Science* 27(12): 1019–1026.

Chen, Y.X., Huang, X.D., Han, Z.Y., Huang, X., Hu, B., Shi, D.Z., and Wu, W.X. 2010b. Effects of bamboo charcoal and bamboo vinegar on nitrogen conservation and heavy metals immobility during pig manure composting. *Chemosphere* 78: 1177–1181.

Chiang, K.Y., Huang, H.J., and Chang, C.N. 2007. Enhancement of heavy metal stabilization by different amendments during sewage sludge composting process. *Journal of Environmental Management* 17(4): 249–256.

Chiroma, T.M., Ebewele, R.O., and Hymore, F.K. 2012. Levels of heavy metals (Cu, Zn, Pb, Fe and Cr) in Bushgreen and Roselle irrigated with treated and untreated urban sewage water. *International Research Journal of Environmental Sciences* 1(4): 50–55.

Demoling, L.A., and Bååth, E. 2008. No long-term persistence of bacterial pollution-induced community tolerance in tylosin-polluted soil. *Environmental Science and Technology* 42: 6917–6921.

Devi, P., and Saroha, A.K. 2014. Risk analysis of pyrolyzed biochar made from paper mill effluent treatment plant sludge for bioavailability and eco-toxicity of heavy metals. *Bioresource Technology* 162: 308–315.

Dominguez, J., and Edwards, C.A. 2004. Vermicomposting organic wastes: A review, in S.H.S. Hanna and W.Z.A. Mikhail (eds.) *Soil Zoology for Sustainable Development in the 21st Century*, pp. 369–395, Eigenverlag, Cairo.

Duruibe, J.O., Ogwuegbu, M.O.C., and Egwurugwu, J.N. 2007. Heavy metal pollution and human biotoxic effects. *International Journal of Physical Sciences* 2(5): 112–118.

Erdem, E., Karapinar, N., and Donat, R. 2004. The removal of heavy metal cations by natural zeolites. *Journal of Colloidal and Interference Science* 280: 309–314.

Fang, M., and Wong, J.W. 1999. Effects of lime amendment on availability of heavy metals and maturation in sewage sludge composting. *Environmental Pollution* 106: 83–89.

Fang, M., and Wong, J.W.C. 2000. Changes in thermophilic bacteria population and diversity during composting of coal fly ash and sewage sludge. *Water, Air and Soil Pollution* 124: 333–343.

Fu, J., Zhao, C., Luo, Y., Liu, C., Kyzas, G.Z., Luo, Y., Zhao, D., An, S., and Zhu, H. 2014. Heavymetals in surface sediments of the Jialu River, China: Their relations to environmental factors. *Journal of Hazardous Materials* 270: 102–109.

Gadepalle, V.P., Ouki, S.K., Herwijnen, R.V., and Hutchings, T. 2007. Immobilization of heavy metals in soil using natural and waste materials for vegetation establishment on contaminated sites. *Soil and Sediment Contamination* 16: 233–251.

Garcia, C., Moreno, J.L., Hernfindez, T., and Costa, F. 1995. Effect ofcomposting on sewage sludges contaminated with heavy metals. *Bioresource Technology* 53: 13–19.

Gea, T., Barrena, R., Artola, A., and Sánchez, A. 2007. Optimal bulking agent particle size and usage for heat retention and disinfection in domestic wastewater sludge composting. *Waste Management* 27(9): 1108–1116.

Ghyasvand, S., Alikhani, H.A., Ardalan, M.M., Savaghebi, G.R., and Hatami, S. 2008. Effect of pre-thermocomposting on decrease of cadmium and lead pollution in vermicomposting of municipal solid waste by *Eisenia fetida*. *American-Eurasian Journal of Agricultural and Environmental Sciences* 4(5): 537–540.

Goswami, L., Sarkar, S., Mukherjee, S., Das, S., Barman, S., Raul, P., Bhattacharyya, P., Mandal, N.C., Bhattacharya, S., and Bhattacharya, S.S. 2014. Vermicomposting of tea factory coal ash: Metal accumulation and metallothionein response in *Eisenia fetida* (Savigny) and Lampito mauritii (Kinberg). *Bioresource Technology* 166: 96–102.

Grøn, C. 2007. Organic contaminants from sewage sludge applied to agricultural soils. *Environmental Science and Pollution Research* 14: 53–60.

Guo, X., Gu, J., Gao, H., Qin, Q., Chen, Z., Shao, L., Chen, L., Li, H., Zhang, W., Chen, S., and Liu, J. 2012. Effects of Cu on metabolisms and enzyme activities of microbial communities in the process of composting. *Bioresource Technology* 108: 140–148.

Gupta, R., and Garg, V.K. 2008. Stabilization of primary sludge during vermicomposting. *Journal of Hazardous Materials* 153: 1023–1030.

Gupta, R., Mutiyar, P.K., Rawat, N.K., Saini, M.S., and Garg, V.K., 2007. Development of a water hyacinth based vermireactor using an epigeic earthworm *Eisenia fetida*. *Bioresource Technology* 98: 2605–2610.

Gusiatin, Z.M., and Kulikowska, D. 2014. The usability of the IR, RAC and MRI indices of heavy metal distribution to assess the environmental quality of sewage sludge composts. *Waste Management* 24: 1227–1236.

Hait, S., and Tare, V. 2012. Transformation and availability of nutrients and heavy metals during integrated composting-vermicomposting of sewage sludges. *Ecotoxicology and Environmental Safety* 79: 214–224.

Han, F.X., Banin, A., Kingery, W.L., Triplett, G.B., Zhou, L.X., Zheng, S.J., and Ding, W.X. 2003. New approach to studies of heavy metal redistribution in soil. *Advances in Environmental Research* 8: 113–120.

Haroun, M., Idris, A., and Omar, S. 2009. Analysis of heavy metals during composting of the tannery sludge using physicochemical and spectroscopic techniques. *Journal of Hazardous Materials* 165: 111–119.

Haynes, R.J., Murtaza, G., and Naidu, R. 2009. Inorganic and organic constituents and contaminants of biosolids: Implications for land application. *Advances in Agronomy* 104: 165–267.

He, M., Li, W., Liang, X., Wu, D., and Tian, G. 2009a. Effect of composting process on phytotoxicity and speciation of copper, zinc and lead in sewage sludge and swine manure. *Waste Management* 29: 590–597.

He, M., Tian, G., and Liang, X. 2009b. Phytotoxicity and speciation of copper, zinc and lead during the aerobic composting of sewage sludge. *Journal of Hazardous Materials* 163: 671–677.

He, X., Zhang, Y., Shen, M., Zeng, G., Zhou, M., and Li, M., 2016. Effect of vermicomposting on concentration and speciation of heavy metals in sewage sludge with additive materials. *Bioresource Technology* 218: 867–873.

Hsu, J.H., and Lo, S.L. 2001. Effect of composting on characterization and leaching of copper, manganese, and zinc from swine manure. *Environmental Pollution* 114: 119–127.

Hua, L., Wu, W., Liu, Y., Bride, M.B.M., and Chen, Y. 2009. Reduction of nitrogen loss and Cu and Zn mobility during sludge composting with bamboo charcoal amendment. *Environmental Science and Pollution Research* 16: 1–9.

Huang, H.J., and Yuan, X.Z. 2016. The migration and transformation behaviors of heavy metals during the hydrothermal treatment of sewage sludge. *Bioresource Technology* 200: 991–998.

Huang, G.F., Wong, J.W.C., Wu, Q.T., and Nagar, B.B. 2004. Effect of C/N on composting of pig manure with sawdust. *Waste Management* 24: 805–813.

Huang, D.L., Zeng, G.M., Jiang, X.Y., Feng, C.L., Yu, H.Y., Huang, G.H., and Liu, H.L. 2006. Bioremediation of Pb-contaminated soil by incubating with *Phanerochaete chrysosporium* and straw. *Journal of Hazardous Materials* 134: 268–276.

Iwegbue, C.M.A., Emuh, F.N., Isirimah, N.O., and Egun, A.C. 2007. Fractionation, characterization and speciation of heavy metals in composts and compost-amended soils. *African Journal of Biotechnology* 6(2): 67–78.

Jain, K. Singh, J., Chauhan, L.K.S., Murthy, R.C., and Gupta, S.K. 2004. Modulation of fly ash-induced genotoxicity in Vicia faba by vermicomposting. *Ecotoxicology and Environmental Safety* 59: 89–94.

Kang, J., Zhang, Z., and Wang, J.J. 2011. Influence of humic substances on bioavailability of Cu and Zn during sewage sludge composting. *Bioresource Technology* 102: 8022–8026.

Karaca, A., Cetin, S.C., Turgay, O.C., and Kizilkaya, R. 2010. Effects of heavy metals on soil enzyme activities, in I. Sherameti and A. Varma (eds.) *Soil Heavy Metals*, Soil Biology, Vol. 19, pp. 237–265, Springer-Verlag, Heidelberg.

Kosobucki, P., Kruk, M., and Buszewski, B., 2008. Immobilization of selected heavy metals in sewage sludge by natural zeolites. *Bioresource Technology* 99: 59–72.

Kulikowska, D., and Gusiatin, Z.M. 2015. Sewage sludge composting in a two-stage system: Carbon and nitrogen transformations and potential ecological risk assessment. *Waste Management* 38: 312–320.

Li, L., Xu, Z., Wu, J., and Tian, G. 2010. Bioaccumulation of heavy metals in the earthworm *Eisenia fetida* in relation to bioavailable metal concentrations in pig manure. *Bioresource Technology* 101: 3430–3436.

Li, Y., Liu, B., Zhang, X., Gao, M., and Wang, J. 2015. Effects of Cu exposure on enzyme activities and selection for microbial tolerances during swine-manure composting. *Journal of Hazardous Materials* 283: 512–518.

Liu, J., Xu, X., Huang, D., and Zeng, G. 2009. Transformation behavior of lead fractions during composting of lead-contaminated waste. *Transactions of Nonferrous Metals Society of China* 19: 1377–1382.

Lv, B., Xing, M., and Yang, J. 2016. Speciation and transformation of heavy metals during vermicomposting of animal manure. *Bioresource Technology* 209: 397–401.

Malley, C., Nair, J., and Ho, G. 2006. Impact of heavy metals on enzymatic activity of substrate and on composting worms *Eisenia foetida. Bioresource Technology* 97: 1498–1502.

Maanan, M., Saddik, M., Maanan, M., Chaibi, M., Assobhei, O., and Zourarah, B. 2015. Environmental and ecological risk assessment of heavy metals in sediments of Nador lagoon. Moroc. *Ecological Indicators* 48: 616–626.

Miretzky, P., Avendaño, M.R., Muñoz, C., and Carrillo-Chavez, A. 2011. Use of partition and redistribution indexes for heavy metal soil distribution after contamination with a multi element solution. *Journal of Soil and Sediments* 11: 619–627.

Nair, A., Juwarkar, A.A., and Devotta, S. 2008. Study of speciation of metals in an industrial sludge and evaluation of metal chelators for their removal. *Journal of Hazardous Materials* 152: 545–553.

Ochoa-Herrera, V., León, G., Banihani, Q., Field, J.A., and Sierra-Alvarez, R. 2011. Toxicity of copper (II) ions to microorganisms in biological wastewater treatment systems. *Science of the Total Environment* 412: 380–385.

Ozcan, S., Tor, A., and Aydin, M.E. 2013. Investigation on the levels of heavy metals, polycyclic aromatic hydrocarbons, and polychlorinated biphenyls in sewage sludge samples and ecotoxicological testing. *Clean Soil Air Water* 41: 411–418.

Pardo, T., Clemente, R., and Bernal, M.P. 2011. Effects of compost, pig slurry and lime on trace element solubility and toxicity in two soils differently affected by mining activities. *Chemosphere* 84: 642–650.

Pathak, A., Dastida, M.G., and Sreekrishnan, T.R. 2009. Bioleaching of heavy metals from sewage sludge: A review. *Journal of Environmental Management* 90: 2343–2353.

Philippis, R.D., and Micheletti, E. 2009. Heavy metals with exopolysaccharide-producing cyanobacteria, in L.K. Wang, J.P. Chen, Y.T. Hung and N.K. Shammas (eds.) *Heavy Metals in the Environment*, pp. 89–122, CRC Press Taylor and Francis Groups, Boca Raton, FL.

Prasad, R., Singh, J., and Kalamdhad, A.S. 2013. Assessment of nutrients and stability parameters during composting of water hyacinth mixed with cattle manure and sawdust. *Research Journal of Chemical Sciences* 3(4): 1–4.

Qiao, L., and Ho, G. 1997. The effects of clay amendment and composting on metal speciation in digested sludge. *Water Research* 31(5): 951–964.

Raichura, A., and McCartney, D., 2006. Composting of municipal biosolids: Effect of bulking agent particle size on operating performance. *Environmental Engineering Science* 5(3): 235–241.

Rohwerder, T., Gehrke, T., Kinzler, K., and Sand, W. 2003. Bioleaching review part A: progress in bioleaching: Fundamentals and mechanisms of bacterial metal sulfide oxidation. *Applied Microbiology and Biotechnology* 63: 239–248.

Rorat, A., Wloka, D., Grobelak, A., Grosser, A., Sosnecka, A., Milczarek, M., Jelonek, P., Vandenbulcke, F., and Kacprzak, M. 2017. Vermiremediation of polycyclic aromatic hydrocarbons and heavy metals in sewage sludge composting process. *Journal of Environmental Management* 187: 347–353.

Samaras, P., Papadimitriou, C.A., Haritou, I., and Zouboulis, A.I. 2008. Investigation of sewage sludge stabilization potential by the addition of fly ash and lime. *Journal of Hazardous Materials* 154: 1052–1059.

Shen, Y., Chen, T.B., Gao, D., Zheng, G.D., Liu, H.T., and Yang, Q.W. 2012. Online monitoring of volatile organic compound production and emission during sewage sludge composting. *Bioresource Technology* 123: 463–470.

Shukla, O.P., Rai, U.N., and Dubey, S. 2009. Involvement and interaction of microbial communities in the transformation and stabilization of chromium during the composting of tannery effluent treated biomass of Vallisneria spiralis L. *Bioresource Technology* 100: 2198–2203.

Siloniz, M.I., Balsalobre, L., Alba, C., and Valderrama, M.J. 2002. Feasibility of copper uptake by the yeast *Pichia guilliermondii* isolated from sewage sludge. *Microbiological Research* 153: 173–180.

Singh, R.P., and Agrawal, M. 2008. Potential benefits and risks of land application of sewage sludge. *Waste Management* 28: 347–358.

Singh, J., and Kalamdhad, A.S. 2011. Effects of heavy metals on soil, plants, human health and aquatic life. *International Journal of Research in Chemistry and Environment* 1(2): 15–21.

Singh, J., and Kalamdhad, A.S. 2012. Concentration and speciation of heavy metals during water hyacinth composting. *Bioresource Technology* 124: 169–179.

Singh, J., and Kalamdhad, A.S. 2013a. Bioavailability and leachability of heavy metals during water hyacinth composting. *Chemical Speciation and Bioavailability* 25(1): 1–14.

Singh, J., and Kalamdhad, A.S. 2013b. Effect of lime on bioavailability and leachability of heavy metals during agitated pile composting of water hyacinth. *Bioresource Technology* 138: 148–155.

Singh, J., and Kalamdhad, A.S. 2013c. Reduction of bioavailability and leachability of heavy metals during vermicomposting of water hyacinth. *Environmental Science and Pollution Research* 20(12): 8974–8985.

Singh, J., and Kalamdhad, A.S. 2013d. Effect of *Eisenia fetida* on speciation of heavy metals during vermicomposting of water hyacinth. *Ecological Engineering* 60: 214–223.

Singh J., and Kalamdhad, A. S. 2014a. Influences of natural zeolite on speciation of heavy metals during rotary drum composting of green waste. *Chemical Speciation and Bioavailability* 26(2): 65–75.

Singh, J., and Kalamdhad, A.S. 2014b. Uptake of heavy metals by natural zeolite during agitated pile composting of water hyacinth. *International Journal of Environmental Science* 5(1): 1–15.

Singh, J., and Kalamdhad, A.S. 2014c. Effects of natural zeolite on speciation of heavy metals during agitated pile composting of water hyacinth. *International Journal of Recycling of Organic Waste in Agriculture* 3(55): 1–17

Singh, J., and Kalamdhad, A.S. 2014d. Effects of carbide sludge (lime) on bioavailability and leachability of heavy metals during rotary drum composting of water hyacinth. *Chemical Speciation and Bioavailability* 26(2): 76–84.

Singh, J., and Kalamdhad, A.S. 2016. Effect of lime on speciation of heavy metals during agitated pile composting of water hyacinth. *Frontiers of Environmental Science and Engineering* 10(1): 93–102.

Singh, J., Prasad, R., and Kalamdhad, A.S. 2013. Effect of natural zeolite on bioavailability and leachability of heavy metals during rotary drum composting of water hyacinth. *Research Journal of Chemistry and Environment* 17(8): 26–34.

Singh, W.R., Pankaj S.K., Singh, J., and Kalamdhad, A.S. 2014. Reduction of bioavailability of heavy metals during vermicomposting of phumdi biomass of Loktak Lake (India) using *Eisenia fetida*. *Chemical Speciation and Bioavailability* 26(3): 158–168.

Singh, J., Kalamdhad, A.S., and Lee, B.K. 2015. Reduction of eco-toxicity risk of heavy metals in the rotary drum composting of water hyacinth: Waste lime application and mechanisms. *Environmental Engineering Research* 20(3): 212–222.

Sizmur, T., and Hodson, M.E. 2009. Do earthworms impact metal mobility and availability in soil?—A review. *Environmental Pollution* 157: 1981–1989.

Song, X.C., Liu, M.Q., Wu, D., Qi, L., Ye, C.L., Jiao, J.G., and Hu, F. 2014. Heavy metal and nutrient changes during vermicomposting animal manure spiked with mushroom residues. *Waste Management* 34: 1977–1983.

Sprynskyy, M., Kosobucki, P., Kowalkowski, T., and Buszewsk, B. 2007. Influence of clinoptilolite rock on chemical speciation of selected heavy metals in sewage sludge. *Journal of Hazardous Materials* 149: 310–316.

Stylianou, M.A., Inglezakis, V.J., Moustakas, K.G., and Loizidou, M.D. 2008. Improvement of the quality of sewage sludge compost by adding natural clinoptilolite. *Desalination* 224: 240–249.

Sundaray, S.K., Nayak, B.B., Lin, S., and Bhatta, D. 2011. Geochemical speciation and risk assessment of heavy metals in the river estuarine sediments-a case study: Mahanadi Basin, India. *Journal of Hazardous Materials* 186: 1837–1846.

Suthar, S. 2008. Bioremediation of aerobically treated distillery sludge mixed with cow dung by using an epigeic earthworm *Eisenia fetida*. *Environmentalist* 28(2): 76–84.

Suthar, S. 2009. Vermistabilization of municipal sewage sludge amended with sugarcane trash using epigeic *Eisenia fetida* (Oligochaeta). *Journal of Hazardous Materials* 163: 199–206.

Suthar, S., and Singh, S. 2008. Feasibility of vermicomposting in biostabilization of sludge from a distillery industry. *Science of the Total Environment* 394: 237–243.

Suthar, S., and Singh, S., 2009. Bioconcentrations of metals (Fe, Cu, Zn, Pb) in earthworm (*Eisenia fetida*), inoculated in municipal sewage sludge: Is earthworm pose a possible risk of terrestrial food-chain contamination? *Environmental Toxicology* 24: 25–32.

Talbot, V.L. 2006. The chemical forms and plant availability of copper in composting organic wastes. Ph D thesis, University of Wolverhampton, Wolverhampton, UK.

Tandy, S., Healey, J.R., Nason, M.A., Williamson, J.C., and Jones, D.L. 2009. Heavy metal fractionation during the co-composting of biosolids, deinking paper fibre and green waste. *Bioresource Technology* 100: 4220–4226.

Vig, A.P., Singh, J., Wani, S.H., and Dhaliwal, S.S. 2011. Vermicomposting of tannery sludge mixed with cattle dung into valuable manure using earthworm *Eisenia fetida* (Savigny). *Bioresource Technology* 102(17): 7941–7945.

Villasenor, J., Rodriguez, L., and Fernandez, F.J. 2011. Composting domestic sewage sludge with natural zeolites in a rotary drum reactor. *Bioresource Technology* 102(2): 1447–1454.

Wang, X., Chen, L., Xia, S., and Zhao, J. 2008. Changes of Cu, Zn, and Ni chemical speciation in sewage sludge co-composted with sodium sulfide and lime. *Journal of Environmental Sciences* 20: 156–160.

Wang, L.M., Zhang, Y.M., Lian, J.J., Chao, J.Y., Gao, Y.X., Yang, F., and Zhang, L.Y. 2013. Impact of fly ash and phosphatic rock on metal stabilization and bioavailability during sewage sludge vermicomposting. *Bioresource Technology* 136: 281–287.

Wong, J.W.C., and Selvam, A. 2006. Speciation of heavy metals during co-composting of sewage sludge with lime. *Chemosphere* 63: 980–986.

Wu, S.H., Shen, Z.Q., Yang, C.P., Zhou, Y.X., Li, X., Zeng, G.M., Ai, S.J., and He, H.J. 2017. Effects of C/N ratio and bulking agent on speciation of Zn and Cu and enzymatic activity during pig manure composting. *International Biodeterioration and Biodegradation* 119: 429–436.

Yan, Y., Zhang, L., Feng, L., Sun, D., and Dang, Y. 2018. Comparison of varying operating parameters on heavy metals ecological risk during anaerobic co-digestion of chicken manure and corn stover. *Bioresource Technology* 247: 660–668.

Yang, J., Zhao, C.H., Xing, M.Y., and Lin, Y.N. 2013. Enhancement stabilization of heavy metals (Zn, Pb, Cr and Cu) during vermifiltration of liquid-state sludge. *Bioresource Technology* 146: 649–655.

Yang, J., Lei, M., Chen, T.B., Gao, D., Zheng, G.D., Guo, G.H., and Le, D.J. 2014. Current status and developing trends of the contents of heavy metals in sewage sludges in China. *Frontiers of Environmental Science and Engineering* 8: 719–728.

Yuan, X., Huang, H., Zeng, G., Li, H., Wang, J., Zhou, C., Zhu, H., Pei, X., Liu, Z., and Liu, Z. 2011. Total concentrations and chemical speciation of heavy metals in liquefaction residues of sewage sludge. *Bioresource Technology* 102: 4104–4110.

Zeng, G.M., Huang, D.L., Huang, G.H., Hu, T.J., Jiang, X.Y., Feng, C.L., Chen Y.N., Tang L., and Liu, H.L. 2007. Composting of lead-contaminated solid waste with inocula of white-rot fungus. *Bioresource Technology* 98: 320–326.

Zhai, Y., Liu, X., Chen, H., Xu, B., Zhu, L., Li, C., and Zeng, G. 2014. Source identification and potential ecological risk assessment of heavy metals in PM2.5 from Changsha. *Science of the Total Environment* 493: 109–115.

Zhu, R., Wu, M., and Yang, J. 2011. Mobilities and leachabilities of heavy metals in sludge with humus soil. *Journal of Environmental Sciences* 23: 247–254.

Zhu, H.N., Yuan, X.Z., Zeng, G.M., Jiang, M., Liang, J., Zhang, C., Yin, J., Huang, H.J., Liu, Z.F., and Jiang, H.W. 2012. Ecological risk assessment of heavy metals in sediments of Xiawan Port based on modified potential ecological risk index. *Transactions of Nonferrous Metals Society of China* 22: 1470–1477.

Zorpas, A.A., and Loizidou, M. 2008. Sawdust and natural zeolite as a bulking agent for improving quality of a composting product from anaerobically stabilized sewage sludge. *Bioresource Technology* 99(16): 7545–7552.

Zorpas, A.A., Constantinides, T., Vlyssides, A.G., Haralambous, I., and Loizidou, M. 2000. Heavy metal uptake by natural zeolite and metals partitioning in sewage sludge compost. *Bioresource Technology* 72: 113–119.

Zorpas, A.A., Vassilis, I., Loizidou, M., and Grigoropoulou, H. 2002. Particle size effects on uptake of heavy metals from sewage sludge compost using natural zeolite clinoptilolite. *Journal of Colloid and Interface Science* 250: 1–4.

Zorpas, A.A., Arapoglou, D., and Panagiotis, K. 2003. Waste paper and clinoptilolite as a bulking material with dewatered anaerobically stabilized primary sewage sludge (DASPSS) for compost production. *Waste Management* 23: 27–35.

5 Bioavailability of Heavy Metals during Composting

5.1 BIOAVAILABILITY OF HEAVY METALS

It is well known that the mobility and bioavailability of heavy metals depends not only on their total concentration but also on their various forms. The study on bioavailability of heavy metals provides more important evidence in determining the toxicity of metals (Nair et al., 2008). A repeating application of compost having high concentration of metals to land can increase metals concentration in the soil even if the heavy-metal concentrations in composting are far below the regulation standard (Chiang et al., 2007). Yobouet et al. (2010) reported that metals in soils can be classified into two fractions: (i) inert fraction and (ii) labile fraction. The inert fraction is supposed to be the nontoxic form, whereas the labile fraction is supposed to be the potentially toxic form (Yobouet et al., 2010). Only the soil labile fraction is taken into account because this fraction is generally called the bioavailable fraction. However, the bioavailable fraction can be changed from one form to another from as well as one receptor to another. The accessibility of metals for plants and microorganisms in the soil system depends mainly on the composition of the various components of soil such as metal bound carbonates, (oxy) metal hydroxides, organic fraction, and silica (Yobouet et al., 2010).

Gupta and Sinha (2007) reported that the bioavailability of metals is considered one of the most critical issues in agricultural studies. Yobouet et al. (2010) stated that the mobilization of pollutants depends on the following factors: mobility of metals, concentration of metals in the soil, and solubility of metals. The solubility of metals depends on the chemical composition of leachate in soil medium; this chemical composition is further changed by the changing pH value that allows metals to exist in their ionic forms (Yobouet et al., 2010). At higher pH, the predominant forms are hydroxides with low solubility, whereas at lower pH the predominant forms are free metallic ions with high solubility. There are two types of metal complexes formed in soil medium: soluble complexes and insoluble complexes. The formation of organometallic complexes of soluble metals can lead to a temporary increase in the solubility of Cu (Yobouet et al., 2010) with pH 9. Figure 5.1 represents the correlation between pH, bioavailability, and solubility of heavy metals.

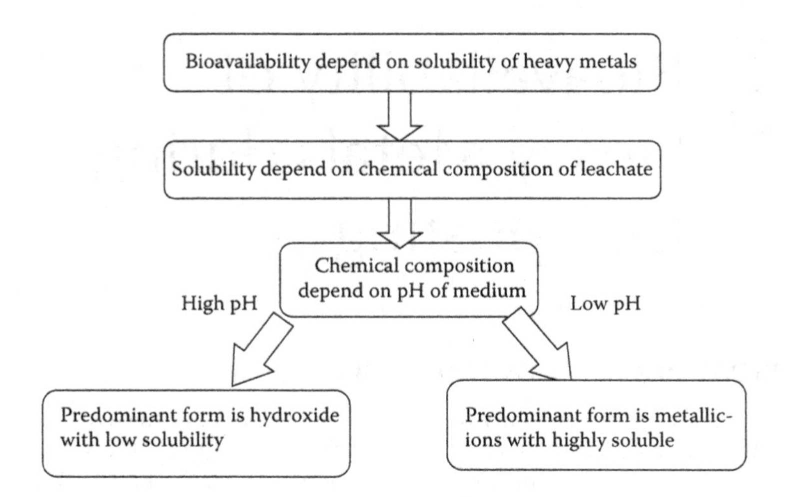

FIGURE 5.1 Correlation among pH, bioavailability, and solubility of heavy metals.

5.1.1 Effects of Physicochemical Parameters on the Bioavailability of Heavy Metals

Physicochemical parameters, such as pH, moisture, temperature, organic matter etc., have a great impact on immobilization of heavy metals during the composting process of different organic wastes. The fractionation of heavy metals in the composting process can be influenced by their discharge through organic matter mineralization or metal solubilization due to a decrease in pH or due to biosorption of metals through the microbial biomass present in the composting mass. Metals form complex with the humic substances during the composting process resulting the metals are insoluble and become difficult to hard for extraction (Castaldi et al., 2006; Cai et al., 2007). The stability constant of metal–humic complexes depends not only on the properties of organics and metals but also it depends on ion strength and pH of the medium (Liu et al., 2008). According to Talbot (2006) changes in pH of the compost, the COOH groups are dissociated and also affect hydrolysis of metal ions. An increases in hydrolysis of metals due to the increases of pH. The Mn and Zn can form complex with organic matter or humic substances at the end of composting process. In the initial phase of the composting process, the neutral pH values of the composting biomass caused a weak adsorption of Cu and Zn onto organic matter due to the leaching of metals under acidic conditions (Lazzari et al., 2000). According to Nomeda et al. (2008), the organometallic complexes follow the Irwing–William series as follows: Cu > Pb > Zn > Cd > Fe.

Electrical conductivity during the final composting should not exceed the salinity limit value of 3 mS/cm to be used as oil amendment (Soumaré, 2003). The complex formation capacity of a humic acid with heavy metals was significantly decreased with increasing electrical conductivity (Talbot, 2006). Chromium (III) has a stronger electrostatic attraction for the sorption sites as compared to the divalent cations, resulting chromium forms the most stable complex with humic substances (Qiao and Ho, 1997). The influence of pH on metal solubility is well known either by solubility equilibria or due to complexation by soluble surface ligands. Therefore, the pH value of the medium increases with a decrease in the solubility and bioavailability of divalent trace metals (Cambier and Charlatchka, 1999). The pH of the composting mixture also affected the solubility of the metal hydroxides and carbonates, whereas the acidic pH increased the solubility of heavy metals due to the higher leachability of metals (Qiao and Ho, 1997). At alkaline pH ($pH > 8$), Ni and Cd can be extracted due to the breakdown of the metal humic complex (Lazzari et al., 2000). Pb is usually considered a small mobile heavy metal in the reducing as well as nonacidic condition. (Lazzari et al., 2000). Organic matter can be decomposed and transformed into stable humic substances, has an ability to interact with toxic metal ions, and has the capability to maintain buffering capacity of compost (Amir et al., 2005). Wu et al. (2017) concluded that the mobility fractions of Cu and Zn were decreased by the composting mixture prepared with initial C/N ratio of 25 during the composting of pig manure. The residual fractions of Cu were found to be dominant in the final composting process. This study also concluded that the C/N ratio of 25 could affect urease activity by impelling the concentration of metal ions and stated that the final compost mixtures prepared with initial C/N ratio of 25 had low risk of Zn and Cu, and this compost can be used in agricultural soils (Wu et al., 2017).

5.1.2 Water-Soluble Metals during the Composting Process

The extraction of metals by distilled water is one of the simplest methods for the evaluation of the bioavailability of heavy metals in the compost and vermicomposting process. Water soluble fraction is easily available to plant, causing possible toxic effect on plant and human health being through consumption of food. It has been suggested that the water-soluble fraction is definitely the most biologically bound fraction. This fraction is not only involved in the contamination of the food chain but also in the contamination of surface water and ground water (Iwegbue et al., 2007). The soil pH controls many of the essential variables in the soil environment such as ion exchange, reduction/oxidation, adsorption, and complexation of reactions. At alkaline pH, adsorptions of cations can occur on the surface of organic matter (Samuel et al., 2013). The effect of the addition of organic material on the solubility of heavy metals depends significantly upon the degree of humification (Gupta and Sinha, 2007). In a risk assessment, the pH is an essential factor that mainly affects the mobility and bioavailability of metals.

Ni is known as a possibly mobile and water-soluble metal and the labile pool in the composting. It is comparatively easily extracted by mild extractants due to its weaker binding to the matrix in comparison to other metals (Smith, 2009). Reduction of Ni during the composting process may be due to the oxidation of the organic fraction and the formation of complex compounds of organic material with Ni (Fang and Wong, 1999).

Hargreaves et al. (2008) reported that the fractions of water-soluble metals (such as Zn, Pb, Cu, and Cd) were decreased and immobilized in the thermophilic stage of composting. In the initial phase of degradation, pH of composting mixture get acidic due to degradation of OM, which triggered availability of Pb and Zn, these metals had the highest water-extractable concentrations in acidic condition (Hargreaves et al., 2008). The increases in total metal concentrations (sample digested with combination of strong acids) during composting was not attended to rise in the water solubility of Cu, Mn and Zn. Water solubility of Cu, Mn and Zn are depend on solubility of organic carbon which increased during the thermophilic phase and then decreased significantly (Hsu and Lo, 2001). Fuentes et al. (2004) reported that the fraction of water-soluble Pb was reduced <0.2 mg/kg, whereas the fraction of water-soluble Ni was found to be <1 mg/kg during the composting of sewage sludge. Hsu and Lo (2001) reported that the fraction of water-soluble Cu increased approximately 16% during the thermophilic phase of composting, and then gradually decreased in the composting of pig manure. The fractions of water-soluble Mn and Zn were increased up to 2% during the thermophilic phase, whereas these fractions were decreased up to 0.5% after the composting process. From this study, it can be concluded that that composting significantly adjusts Cu extractability in the pig manure and seems to have a minor impact on bioavailability of Mn or Zn. Cr is considered relatively stable and its stability cannot be changed in simple environmental conditions (Haroun et al., 2009). However, Cr shows amphotericity, i.e., the ability of Cr to react with acid and base both, subsequently its solubility and potential mobility increased in the environment (Haroun et al., 2009).

Ciavatta et al. (1993) reported that Cr tends to insolubilize during the composting process, and Cr released from organic matter in neutral or alkaline soils precipitates as insoluble forms; hence, Cr is not accumulated by plants. The bioavailability of Cr is limited due to its low solubility, resulting Cr does not show phytotoxicity in plants. The fractions of water-soluble metals were increased due to decomposition of organic matter in the thermophilic stage of composting, whereas these fractions were decreased in the final composting process (Ahmed et al., 2007). Table 5.1 shows the changes in water-soluble Zn, Cu, Mn, Fe, and Cr concentrations during the agitated pile composting of water hyacinth in the rotary drum reactor. Singh and Kalamdhad (2013a) reported that the fractions of heavy metals were decreased during the agitated pile composting of water hyacinth. The water-solubility fractions of Zn, Cu, Mn, Fe, and Cr were decreased (% of total metal) from 2.1% to 1.5%, from 3.6% to 2.2%, from 3.2% to 1.2%, and from 0.8% to 0.2%, respectively, whereas the water-solubility fraction of Fe was increased from 0.2% to 0.6% at the end of the composting period.

TABLE 5.1
Changes in Water Solubility of Heavy Metals during Water Hyacinth Composting

Composting Methods	Days	Water-soluble Metals Concentration (mg/kg)							
		Zn	Cu	Mn	Fe	Ni	Pb	Cd	Cr
Agitated pile	0	3.4	1.1	18.5	16.14	ND	ND	ND	2.0
	30	4.5	2.3	13.2	75.1	ND	ND	ND	0.54
Rotary drum/reactor	0	2.3	2.3	10	19.3	ND	ND	ND	1.7
	20	1.2	1.1	3.4	20.3	ND	ND	ND	0.4

Note: Composition of compost materials water hyacinth (90 kg), sawdust (15 kg), and cattle manure (45 kg).
ND, not detected.

Singh and Kalamdhad (2013b) reported the reduction of water-soluble metals in the water hyacinth composting in the rotary drum reactor. The water-soluble concentrations of Zn, Cu, Mn, Fe, and Cr were decreased (% of total metals) from 1.5% to 0.8%, from 2.8% to 1.2%, from 1.9% to 0.5%, from 0.3% to 0.2%, and 2.9% to 0.5% respectively, during the composting process. Hsu and Lo (2001) also reported that the concentration of water-soluble Cu was decreased up to 3% after the composting of pig manure. The fractions of water-soluble Ni, Pb, and Cd were not observed in the all trials after the composting process of water hyacinth (Singh and Kalamdhad, 2013a,b). Castaldi et al. (2006) reported that the fractions of water-soluble Pb and Cd were found to be approximately 0.89 and 0.005 mg/kg d.m., respectively, after the composting of municipal solid waste. Sims and Kline (1991) reported the water-extractable Ni fraction to be approximately 7.2% of total Ni during the composting of sewage sludge. The sequence of water-soluble metal concentration in the composting of water hyacinth in the rotary drum reactor was as follows: Fe > Mn > Cu > Zn > Cr (Singh and Kalamdhad, 2013b). The sequence of water-soluble metal concentrations in the agitated pile composting water hyacinth was as follows: Fe > Mn > Zn > Cu > Cr (Singh and Kalamdhad, 2013a).

Singh et al. (2015) reported that the fractions of the water solubility of heavy metals (e.g., Zn, Cu, Mn, Fe, Pb, and Cr) were decreased significantly in all trials during the composting of *Salvinia natans* weed. The sequence of reduction of water-soluble heavy metals was as follows: Fe (51.1%) > Pb (39.7%) > Mn (39.5%) > Cu (38.4%) > Cr (28.3%) > Zn (25.4%). Singh et al. (2016) studied the variations in the water-soluble Zn, Mn, Cu, Fe, Pb, and Cr contents during the 20 day composting period of water fern. The water solubility of heavy metals was decreased significantly at the end of the composting process. However, the highest proportions of the water-solubility reduction of Zn, Cu, Mn, Fe, Pb and Cr were found to be associated with 43.4%, 58.4%, 60.7%, 70.6%, 54.7% and 62.0% of total fractions of these metals. Results of this study confirmed that the addition of cattle manure and rice husk provides an adequate amount of nutrients to the composting microorganisms for the fast and high degradation of organic matter, water solubility of heavy metals decreased at the final stage of composting due to binding of metals with humic substances present in the compost (Singh et al., 2016).

Singh and Kalamdhad (2016) reported that water solubility of metals was decreased approximately 27.5% for Zn, 35.1% for Cu, 36.3% for Mn, 37.8% for Fe, 80.0% for Pb, and 50.6% for Cr during the vermicomposting process of *S. natans*. The sequence of water-soluble heavy metals at the end of the vermicomposting process was as follows: Fe > Mn > Cr > Zn > Cu > Pb. The water-soluble fractions of Ni and Cd were not observed during the vermicomposting process of *S. natans*. Meng et al. (2017) studied water solubility of Cu and Zn during the composting of pig manure, and reported that the water-soluble fractions of Cu and Zn decreased gradually from 115.11 to 31.37 mg/kg for Cu and from 59.99 to 21.95 mg/kg for Zn at the end of the composting period of 84 days. The water-soluble Cu fraction was found to be higher than that of Zn in the mature composting of pig manure. Hazarika et al. (2017) conducted a study on the changes in water solubility of heavy metals during the composting of paper mill sludge. This study also reported that water extractability of Pb, Cr, and Hg was not observed during the composting process. Water extractability of Mn, Cd, and Cu was decreased significantly at the end of the composting process. Zn was not detected in trial 3, whereas Fe was not detected in trial 1. The water solubility of Zn was increased in trial 5, whereas the water solubility of Fe was increased in trial 3. Vishan et al. (2017) studied the changes in water-soluble fractions of Pb, Zn, Ni, and Cd during the composting process in the rotary drum reactor. This study reported that the water solubility of Zn was decreased by 41% in the final composting process, while the water solubility of Ni, Cd, and Pb was not detected during the composting process.

Singh and Kalamdhad (2012) reported that the fraction of heavy metals was increased moderately with the degradation of organic matter, whereas the water solubility of metals was decreased due to the changes of ionic and oxidizing conditions of the composting environment at the end of the composting process.

5.1.3 Diethylene Triamine Pentracetic Acid (DTPA)-Extractable Heavy Metals

DTPA is a chelating compound and has been widely used to assess the plant uptake availability of the metals in soil/compost (Guan et al., 2011). The DTPA extractability of metals may characterize a complementary methodology to check the availability of metals in the soil and sludge-amended soil to plant uptake (Fang and Wong, 1999; Fuentes et al., 2006). It has been suggested that the DTPA solution can extract both carbonate-bound and organically bound metal fractions available in calcareous soils, and specify the concentration of metals possibly available for plant growth (Walter et al., 2006). The processes involved in metal uptake followed by accumulation by various plants depend on the concentration of availability of metals, solubility progressions, and plant growth species (Gupta and Sinha, 2007). The fate of toxic metals mainly depends on their relations with inorganic and organic components of soil surfaces (Bragato et al., 1998). Bragato et al. (1998) reported that the DTPA solution extracted approximately 6% for Zn, 3% for Ni, 29% for Cu, and 24% for Pb (% of their total content) in control soil. DTPA extractability of Cu and Pb was not changed during the process, whereas DTPA extractability of Zn and Ni was increased significantly in the composting of sewage sludge. This study also reported that DTPA-extractable Zn was found to be approximately 9%, whereas DTPA-extractable Ni was found to be approximately 4% of their total concentration in the soil. DTPA extractability of Zn and Ni was found to be 11% and 5%, respectively. Wong and Selvam (2006) stated the DTPA-extractable concentration of Cu, Mn, Ni, and Pb and Zn contents was decreased with an increase in lime amendment rate. The changes of Ni and Zn concentration among control and lime treatments were not observed significantly. The sequence of metal contents in DTPA extracts in the sludge composting was as follows: Zn > Cu > Mn > Pb > Ni, whereas the sequence of metal contents in DTPA extracts at the end of the composting process was as follows: Zn > Mn > Cu > Pb > Ni.

Table 5.2 shows the changes in extraction efficiency of DTPA-extractable Zn, Cu, Mn, Fe, Ni, and Cr during the composting of water hyacinth. DTPA-extractable Pb and Cd were not observed in the composting of water hyacinth. Singh and Kalamdhad (2013a) reported the reduction of DTPA-extractable metals during the

TABLE 5.2
Changes in DTPA Extractability of Heavy Metals during Water Hyacinth Composting

Composting Methods	Days	DTPA-extractable Heavy Metals Concentration (mg/kg)							
		Zn	Cu	Mn	Fe	Ni	Pb	Cd	Cr
Agitated pile	0	31.8	11.6	151.3	187.9	3.96	ND	ND	7.4
	30	57.4	5.6	53.4	334.9	1.785	ND	ND	5.5
Rotary drum/reactor	0	30.8	15.9	230.5	137.3	1.57	ND	ND	5.4
	20	51.7	6.9	213.7	228.1	0.675	ND	ND	2.7

Note: Composition of compost materials water hyacinth (90 kg), sawdust (15 kg), and cattle manure (45 kg).
ND, not detected.

agitated pile composting of water hyacinth. The DTPA extraction of Zn, Cu, Mn, Fe, Ni, and Cr was decreased (% of total metal) approximately 39.87%, 81.4%, 81.7%, 60.7%, 64.2%, and 31.1%, respectively, in the composting process. The sequence of DTPA-extractable heavy-metal contents in the final composting of water hyacinth was as follows: Fe > Mn > Zn > Ni > Cu > Cr (Singh and Kalamdhad, 2013a). Singh and Kalamdhad (2013b) revealed that the DTPA-extractable Zn and Fe contents were increased due to the dissolution of organic matter during the DTPA extraction procedure in the composting of water hyacinth in the rotary drum reactor. However, the concentrations of Cu, Mn, Ni, and Cr were decreased approximately 56.8%, 7.3%, 57.0%, and 50.7% in the final composting. Singh and Kalamdhad (2013b) stated that the DTPA-extractable Fe was <5% of total Fe in the final composting of water hyacinth, whereas Walter et al. (2006) reported approximately 3%–9% of total Fe in the composting of sewage sludge. DTPA-extractable Pb and Cd contents were not observed during the composting of water hyacinth. The total Cu concentration was very low, whereas it was higher in plant uptake in the final composting of water hyacinth (Singh and Kalamdhad, 2013a,b). The stability of metal–humic complexes depends mainly on the composition of organic materials, metal, ion strength, and pH of the soil medium (Liu et al., 2008). The DTPA-extractable Cr concentration was decreased due to low mobility and availability of Cr in insoluble form in the final composting of water hyacinth (Walter et al., 2006), whereas DTPA-extractable heavy metals was decreased due to the formation of metal-humic complexes during the composting process (Liu et al., 2008). Singh et al. (2015) reported that DTPA extractability of heavy metals was decreased significantly during the composting process. The sequence of DTPA reduction was as follows: Cr (56.0%) > Fe (39.5%) > Pb (36.1%) > Cu (32.8%) > Zn (25.3%) > Ni (22.9%) > Mn (22.5%). However, Cd was not detected in the composting of *S. natans*. Singh and Kalamdhad (2016) reported that the DTPA extractability of metals was decreased approximately 31.0% for Zn, 39.2% for Cu, 18.2% for Mn, 37.2% for Fe, 82.7% for Pb, 78.9% for Ni, and 61.9% for Cr during the vermicomposting process. The sequence of DTPA-extractable heavy metals in the vermicomposting was as follows: Fe > Mn > Zn > Cr > Pb > Ni > Cu.

However, Cd was not detected in the vermicomposting of *S. natans*. Singh et al. (2016) reported that the DTPA extraction of Zn, Cu, Mn, Ni, Pb, Fe, and Cr was decreased significantly during the composting of water fern in the rotary drum reactor. This study also reported that the reduction in DTPA extractability of heavy metals was found to be approximately 44.8% for Zn, 65.8% for Cu, 50.4% for Mn, 52.1% for Pb, 43.4% for Ni, 73.0% for Cr, and 56.8% for Fe in the final composting of water fern. Hazarika et al. (2017) reported the DTPA concentration for Zn, Cu, and Mn during composting of paper mill sludge in the rotary drum reactor. The concentration of Cu was increased in trial 1due to incomplete degradation in the composting of the paper mill sludge. Wang et al. (2016) reported that the fractions of DTPA-extractable Cu and Zn were decreased in all treatments with medical stone at the end of composting of pig manure. The addition of medical stone dosages significantly decreased the DTPA extractability of Cu and Zn. However, the DTPA extractability of Zn was increased 3.27% in the control at the end of composting process. Finally, this study revealed that the addition of medical stone into the composting mixtures decreased the bioavailability of Cu and Zn due to the immobilization capacity of heavy metals by medical stone.

Vishan et al. (2017) reported that DTPA extractability of Pb, Ni, and Cd was decreased significantly during the composting period of 20 days, whereas the DTPA extractability Pb and Cd were not observed throughout the 20 day composting period. The concentration of Zn was increased approximately 80% from its initial value. Meng et al. (2017) studied DTPA extraction of Cu and Zn during the composting of pig manure and reported that DTPA Cu and Zn concentrations slightly decreased from 394.0 to 375.8 mg/kg for Cu and from 483.47 to 399.3 mg/kg for Zn at 84 days of composting. These results specified that the potential toxicity threats of Cu and Zn in raw pig manure were found to be more severe as compared to mature composting. From this study, it can be concluded that the composting process could successfully reduce heavy-metal bioavailability.

Awasthi et al. (2016) reported that the DTPA-extractable Cu, Zn, Ni, and Pb contents changed in different treatment amendments with lime and biochar during the composting period. DTPA-extractable Cu and Zn contents were slightly increased in lime and biochar treatments during the biooxidative phase and then slowly reduced; however, in control treatment, contents of Cu and Zn were increased from 28.34 to 46.16 mg/kg and from 270.2 to 485.1 mg/kg, respectively, in the final compost. Bioavailability of Cu and Zn in the both biochar + lime and lime-treated composts was decreased significantly, whereas the bioavailability of these metals in the biochar + lime-added compost was increased. This study also reported that the DTPA-extractable Ni and Pb primarily enhanced and then steadily reduced at the end of the composting period in all treatments. Bioavailability of Ni content was increased from 7.45 to 8.82 mg/kg and 7.48 to 8.68 mg/kg in control and lime treatment, respectively, until day 21, whereas in combined amendment of biochar and lime Ni contents increased only until day 14 and then reduced gradually at the end of composting. Similar to Ni, bioavailability of Pb was detected in all amendments. This study concluded that the combined effect of lime and biochar was highly effective for the immobilization of selected heavy metals (Awasthi et al., 2016).

5.2 CONCLUSION

The composting process could decrease the bioavailability of heavy metals. Toxicity of heavy metals depends not only on the bioavailable fraction but also on the total concentration of metal in the composting process. The pH value has a key role in the bioavailability of metals in the composting of various wastes. The total concentrations of Cu and Zn were very low as compared to the Fe, Cr and Mn whereas the concentrations of water solubility, plant availability (DTPA extractability), and leachability of Fe, Cr, and Mn were found to be higher. The total Cu content was less than the total Cr content but the water solubility of Cu was higher than Cr, indicating that Cu has a higher toxicity risk than Cr. The addition of the optimum amount of cattle manure could enhance organic matter decomposition, resulting humus like substances are formed that bind up water soluble and plant available fractions of metals. Therefore, it reduced the toxicity risk of metals during the composting process. Therefore, the toxicity of metals depends on its water solubility, plant availability, and leachability fraction rather than on the total metal concentration. Water-soluble concentrations of Ni, Pb, and Cd were not evaluated during the composting of water hyacinth. DTPA-extractable Pb and Cd were not detected throughout the composting period. The rotary drum composter was highly efficient for making compost from the different waste materials due to appropriate agitation, mixing, and aeration. The bioavailability of metals depends on physicochemical properties of the medium rather than the total metal contents. The total Cu content was less than the total Zn content but water solubility of Cu was similar to Zn, which indicates the toxicity of both metals. The lowest water solubility of Cr was observed in all treatments in the composting of water hyacinth. The addition of medical stone into the composting mixtures of pig manure decreases the bioavailability of Cu and Zn in the final composting. DTPA extractability of Cu, Zn, Ni, and Pb were reduced in lime and biochar amendments during the sewage sludge composting.

REFERENCES

Ahmed, M., Idris, A., and Omar, S.R.S. 2007. Physicochemical characterization of compost of the industrial tannery sludge. *Journal of Engineering Science and Technology* 2(1): 81–94.

Amir, S., Hafidi, M., Merlina, G., and Revel, J.C. 2005. Sequential extraction of heavy metals during composting of sewage sludge. *Chemosphere* 59: 801–810.

Awasthi, M.K., Wang, Q., Huang, H., Li, R., Shen, F., Lahori, A.H., Wang, P., Guo, D., Guo, Z., Jiang, S., and Zhang, Z. 2016. Effect of biochar amendment on greenhouse gas emission and bio-availability of heavy metals during sewage sludge co-composting. *Journal of Cleaner Production* 135: 829–835.

Bragato, G., Leita, L., Figliolia, A., and Nobili, M. 1998. Effects of sewage sludge pretreatment on microbial biomass and bioavailability of heavy metals. *Soil and Tillage Research* 46: 129–134.

Cai, Q.Y., Mo, C.H., Wu, Q.T., Zeng, Q.Y., and Katsoyiannis, A. 2007. Concentration and speciation of heavy metals in six different sewage sludge-composts. *Journal of Hazardous Materials* 147: 1063–1072.

Cambier, P., and Charlatchka, R. 1999. Influence of reducing conditions on the mobility of divalent trace metals in soils, in H.M. Selim and I.K. Iskandar (eds.) *Fate and Transport of Heavy Metals in the Vadose Zone*. Lewis Publishers, Boca Raton, FL/London/New York.

Castaldi, P., Santona, L., and Melis, P. 2006. Evolution of heavy metals mobility during municipal solid waste composting. *Fresenius Environmental Bulletin* 15(9b): 1133–1140.

Chiang, K.Y., Huang, H.J., and Chang, C.N. 2007. Enhancement of heavy metal stabilization by different amendments during sewage sludge composting process. *The Journal of Environmental Management* 17(4): 249–256.

Ciavatta, C., Govi, M., Simoni, A., and Sequi, P. 1993. Evaluation of heavy metals during stabilization of organic matter in compost produced with municipal solid wastes. *Bioresource Technology* 43: 147–153.

Fang, M., and Wong, J.W.C. 1999. Effects of lime amendment on availability of heavy metals and maturation in sewage sludge composting. *Environmental Pollution* 106: 83–89.

Fuentes, A., Llorens, M., Saez, J., Aguilar, M.I., Soler, A., Ortuno, J.F., and Meseguer, V.F. 2004. Simple and sequential extractions of heavy metals from different sewage sludges. *Chemosphere* 54: 1039–1047.

Fuentes, A., Llorens, M., Saez, J., Aguilar, M.I., Marın, A.B.P., Ortuno, J.F., and Meseguer, V.F. 2006. Ecotoxicity, phytotoxicity and extractability of heavy metals from different stabilised sewage sludges. *Environmental Pollution* 143: 355–360.

Guan, T.X., He, H.B., Zhang, X.D., and Bai, Z. 2011.Cu fractions, mobility and bioavailability in soil-wheat system after Cu-enriched livestock manure applications. *Chemosphere* 82: 215–222.

Gupta, A.K., and Sinha, S. 2007. Phytoextraction capacity of the plants growing on tannery sludge dumping sites. *Bioresource Technology* 98: 1788–1794.

Hargreaves, J.C., Adl, M.S., and Warman, P.R. 2008. A review of the use of composted municipal solid waste in agriculture. *Agriculture Ecosystem and Environment* 123: 1–14.

Haroun, M., Idris, A., and Omar, S. 2009. Analysis of heavy metals during composting of the tannery sludge using physicochemical and spectroscopic techniques. *Journal of Hazardous Materials* 65: 111–119.

Hazarika, J., Ghosh, U., Kalamdhad, A.S., Khwairakpam, M., and Singh, J. 2017. Transformation of elemental toxic metals into immobile fractions in paper mill sludge through rotary drum composting. *Ecological Engineering* 101: 185–192.

Hsu, J.H., and Lo, S.L. 2001. Effects of composting on characterization and leaching of copper, manganese, and zinc from swine manure. *Environmental Pollution* 114: 119–127.

Iwegbue, C.M.A., Emuh, F.N., Isirimah, N.O., and Egun, A.C. 2007. Fractionation, characterization and speciation of heavy metals in composts and compost-amended soils. *African Journal of Biotechnology* 6(2): 67–78.

Lazzari, L., Sperni, L., Bertin, P., and Pavoni, B. 2000. Correlation between inorganic (heavy metals) and organic (PCBs and PAHs) micropollutant concentrations during sewage sludge composting processes. *Chemosphere* 41: 427–435.

Liu, S., Wang, X., Lu, L., Diao, S., and Zhang, J. 2008. Competitive complexation of copper and zinc by sequentially extracted humic substances from manure compost. *Agricultural Sciences in China* 7(10): 1253–1259.

Meng, J., Wang, L., Zhong, L., Liu, X., Brookes, P.C., Xu, J., and Chen, H. 2017. Contrasting effects of composting and pyrolysis on bioavailability and speciation of Cu and Zn in pig manure. *Chemosphere* 180: 93–99.

Nair, A., Juwarkar, A.A., and Devotta, S. 2008. Study of speciation of metals in an industrial sludge and evaluation of metal chelators for their removal. *Journal of Hazardous Materials* 152: 545–553.

Nomeda, S., Valdas, P., Chen, S.Y., and Lin, J.G. 2008. Variations of metal distribution in sewage sludge composting. *Waste Management* 28: 1637–1644.

Qiao, L., and Ho, G. 1997. The effects of clay amendment and composting on metal speciation in digested sludge. *Water Research* 31(5): 951–964.

Samuel, P., Ingmar, P., Boubie, G., and Daniel, L. 2013. Trivalent chromium removal from aqueous solution using raw natural mixed clay from Burkina Faso. *International Research Journal of Environmental Sciences* 2(2): 30–37.

Sims, J.T., and Kline, J.S. 1991. Chemical fractionation and plant uptake of heavy metals in soil emended with co-composted sewage sludge. *Journal of Environmental Quality* 20: 387–395.

Singh, J., and Kalamdhad, A.S. 2012. Concentration and speciation of heavy metals during water hyacinth composting. *Bioresource Technology* 124: 169–179.

Singh, J., and Kalamdhad, A.S. 2013a. Bioavailability and leachability of heavy metals during water hyacinth composting. *Chemical Speciation and Bioavailability* 25(1): 1–14.

Singh, J., and Kalamdhad, A.S. 2013b. Assessment of bioavailability and leachability of heavy metals during rotary drum composting of green waste (water hyacinth). *Ecological Engineering* 52: 59–69.

Singh, W.R., Pankaj, S.K., and Kalamdhad, A.S. 2015. Reduction of bioavailability and leachability of heavy metals during agitated pile composting of *Salvinianatans* weed of Loktak Lake. *International Journal of Recycling of Organic Waste in Agriculture* 4: 143–156.

Singh, W.R., Kalamdhad, A.S., and Singh, J. 2016. The preferential composting of water fern and a reduction of the mobility of potential toxic elements in a rotary drum reactor. *Process Safety and Environmental Protection* 102: 485–494.

Singh, W.R., and Kalamdhad, A.S. 2016. Transformation of nutrients and heavy metals during vermicomposting of the invasive green weed Salvinia natans using *Eisenia fetida. International Journal of Recycling of Organic Waste in Agriculture* 5: 205–220.

Smith, S.R. 2009. A critical review of the bioavailability and impacts of heavy metals in municipal solid waste composts compared to sewage sludge. *Environment International* 35: 142–156.

Soumaré, M., Tack, F.M.G., and Verloo, M.G. 2003. Characterisation of Malian and Belgian solid waste composts with respect to fertility and suitability for land application. *Waste Management* 23: 517–522.

Talbot, V.L. 2006. The chemical forms and plant availability of copper in composting organic wastes. Ph.D. Thesis. University of Wolverhampton, Wolverhampton, UK.

Vishan, I., Sivaprakasam, S., and Kalamdhad, A. 2017. Isolation and identification of bacteria from rotary drum compost of water hyacinth. *International Journal of Recycling of Organic Waste in Agriculture* 6: 245–253.

Walter, I., Martinez, F., and Cala, V. 2006. Heavy metal speciation and phytotoxic effects of three representative sewage sludge for agricultural uses. *Environmental Pollution* 139: 507–514.

Wang, Q., Wang, Z., Awasthi, M.K., Jiang, Y., Li, R., Ren, X., Zhao, J., Shen, F., Wang, M., and Zhang, Z. 2016. Evaluation of medical stone amendment for the reduction of nitrogen loss and bioavailability of heavy metals during pig manure composting. *Bioresource Technology* 220: 297–304.

Wong, J.W.C., and Selvam, A. 2006. Speciation of heavy metals during co-composting of sewage sludge with lime. *Chemosphere* 63: 980–986.

Wu, S.H., Shen, Z.Q., Yang, C.P., Zhou, Y.X., Li, X., Zeng, G.M., Ai, S.J., and He, H.J. 2017. Effects of C/N ratio and bulking agent on speciation of Zn and Cu and enzymatic activity during pig manure composting. *International Biodeterioration and Biodegradation* 119: 429–436.

Yobouet, Y.A., Adouby, K., Trokourey, A., and Yao, B. 2010. Cadmium, copper, lead and zinc speciation in contaminated soils. *International Journal of Engineering, Science and Technology* 2(5): 802–812.

6 Chemical Speciation of Heavy Metals during the Composting Process

6.1 SPECIATION OF HEAVY METALS

The evaluation of total heavy-metal concentration (sample digested with combination of strong acids) in the final compost is not presented suitable information about risk of bioavailability, toxicity, capacity for immobilization of heavy metals in the environment and chemical forms of a metal are accessible in the composting process (Liu et al., 2007). The bioavailability of metals in soil is a dynamic process that depends on specific combinations of chemical, biological, and environmental parameters. The bioavailability, movement of metals, and their eco-toxicity to plants depend mainly on their particular chemical forms or methods of binding (Fuentes et al., 2004). Metals are released from the organic material during its decomposition in the composting process by the action of microorganisms (Qiao and Ho, 1997).

There are various extraction reagents and extraction processes to evaluate the bioavailability of potentially toxic metals (Nemeth et al., 1999). There are both single and sequential extraction (SE) approaches that are mostly applied in the chemical speciation of metals: to explain the chemistry of soil, to analyze the structure and configuration of soil constituents, and to progress consideration of the processes happening in the soil system.

The SE processes deliver convenient information about the risk valuation of toxic metals. Subsequently, the quantity of metals mobilized in various experimental conditions may be evaluated. The SE technique might deliver different form of metals along with their natures and permit the expectation of metal bioavailability (He et al., 2009a). The mobility and bioavailability of the metals were reduced roughly with the sequence of extraction (Nair et al., 2008). In a SE procedure various chemical extracting reagents are applied to a sample successively, to dissolve the desired constituents of the sample matrix. An applied reagent will released all the metals fractions (i.e., exchangeable, carbonate fractions, etc.) from a specified matrix of the sample , one reagent will not affect the component of other component will be used in next step (Li et al., 2001). The main benefit claimed for SE over the practice of single extraction is improvement in specific phases. The water-extractable and exchangeable fractions are considered one of most potentially available fraction of the total heavy metals (Pare et al., 1999).

SE can be used in composts or various wastes to examine the distribution of heavy metals in different fractions. Meanwhile, it can provide a more exact degree of the environmental risk. Pare et al. (1999) stated that the water-soluble and exchangeable fractions of metals are considered the most available fractions of the plants. The chemical speciation may be defined as the procedure of classifying and measuring the amount of various species, fractions, or phases occurring in the sample. Species may be demarcated: (i) functionally, such species may occur due to plant uptakes, (ii) operationally, such species may allow processes or reagents used in the extraction of different fractions, and (iii) especially, such species may be used as exact constituents or oxidation conditions of the metal (Fuentes et al., 2004).

Tessier et al. (1979) reported that the SE of heavy metals can be classified into five fractions): the exchangeable (F1) and carbonate fractions (F2) are the most easily bioavailable fractions and can be separated from the matrix of sample may cause a threat to humans by changing groundwater quality; whereas the reducible (F3) and oxidizable fractions (F4) are not released under normal environmental conditions (Venkateswaran et al., 2007). The heavy metals present in the residual fraction (F5) are generally not predicted to release quickly in natural circumstances (Gupta and Sinha, 2007; Singh and Kalamdhad, 2013). Zheng et al. (2007) reported that the speciation of heavy metals plays a key role in the toxicity of metals occurring in the sewage sludge/compost followed by land use (Zheng et al., 2007). The assessment of the toxicity risk of metal-contaminated sewage sludge or its compost by using SE procedure is very essential to evaluate the appropriateness of sludge or its compost to be used in agricultural lands (Walter et al., 2006). According to Amir et al. (2005), the chemical speciation allows the assessment of heavy-metal bioavailability and also evaluates different characteristics of the metals and their affection strength. Metals may be present in their free ionic form or complexed with organic fraction or combined in the mineral portion of the compost sample (Amir et al., 2005).

Nomeda et al. (2008) reported that the mobility of metals was increased during the initial phase of composting due to acid production/increasing pH. Qiao and Ho (1997) reported that red mud has the potential to change the distribution of heavy metals by enhancing the pH, solid-to-liquid ratio, and availability of adsorption sites. Nomeda et al. (2008) reported that the metals bound with sulfides in a composting mixture were possibly oxidized to form metal sulfates; hence, metals bound with sulfides can be unconfined and then adsorbed on surface of sulfate as well as precipitated in the form of metal oxides.

Venkateswaran et al. (2007) stated that both the exchangeable and carbonate fractions of heavy metals are considered the most mobile fractions and are easily absorbed by plants after leaching when environmental conditions are changed. In addition, these fractions may pose a threat to humans through polluting groundwater. Acidic conditions improve the movement of metal in the surrounding environment; consequently, the metals bound to the carbonate fraction are considered to be highly sensitive to lower pH and leached out (Zheng et al., 2007).

Gupta and Sinha (2007) reported that the residual fraction of metal- is commonly not expected to release in the environment under the normal environmental conditions.

Bioavailability of metals depends on numerous factors that are connected to the nature of the metals, the soil assets (e.g., pH, Eh, clay content, soil organic matter), and the type and quality of soil improvements (Achiba et al., 2009). The mobility of metals in the soils can be evaluated on the basis of their complete and comparative contents of fractions which are weakly bound to the soils (Achiba et al., 2009). Immobilization of metal fractions may be explained by the fact that metal fractions are commonly adsorbed on hydrous oxides of iron and involved in cation exchange reactions (Khaled, 2004). Tables 6.1 and 6.2 present the chemical speciation of heavy metals in various waste materials during the composting process.

Singh and Kalamdhad (2012) reported that the F1, F2, F3, F4, and F5 fractions of different heavy metals (Zn, Cu, Mn, Fe, Ni, Pb, Cd and Cr), their total metal contents of in sawdust (i) cattle manure (ii) water hyacinth (iii) (Singh and Kalamdhad, 2012). The order of different fractions of Zn in various raw materials was as follows: F5 (64%) > F4 (12%) > F1 (10%) > F3 (8%) > F2 (6%) in water hyacinth; F5 (77%) > F1 (10%) > F2 (7%) > F3 (5%) > F4 (2%) in sawdust; and F5 (53%) > F2 (24%) > F3 (10%) > F4 (8%) > F1 (6%) in cattle manure. The residual fractions of Zn were predominant in all three raw materials. The order of different fractions of Cu in different raw materials was as follows: F5 (56%) > F4 (34%) > F1 (5%) > F2 (3%) > F3 (2%) in water hyacinth; F5 (59%) > F4 (30%) > F1 (6%) > F3 (3%) > F2 (2%) in sawdust; and F5 (49%) > F4 (38%) > F1 (7%) > F2 (4%) > F3 (2%) in cattle manure. The order of different fractions of Mn in different raw materials was as follows: F5 (40%) > F1 (28%) > F4 (13%) > F3 (11%) > F2 (8%) in water hyacinth; F5 (57%) > F1 (30%) > F2 (6%) > F3 (5%) > F4 (2%) in sawdust; and F1 (44%) > F5 (24%) > F2 (23%) > F3 (7%) > F4 (3%) in cattle manure. The order of different fractions of Fe in different raw materials was as follows: F4 (55%) > F3 (30%) > F5 (14%) > F1 (1%) > F2 (0.3%) in water hyacinth; F5 (47%) > F3 (30%) > F4(19%) > F2 (2%) > F1 (2%) in sawdust; and F5 (46%) > F4 (26%) > F3 (25%) > F2 (2%) > F1 (1%) in cattle manure. The F3 and F4 fractions of Fe were found to be dominant in the water hyacinth. The order of different fractions of Ni was as follows: F5 (93%) > F1 (6%) > F2 (1%); F5 (98%) > F1 (2%); and F5 (98%) > F1 (2%) in water hyacinth, sawdust, and cattle manure, respectively. The F5 fraction of Ni was found mainly in all three materials (Singh and Kalamdhad, 2012). Fuentes et al. (2004) reported that the residual fraction of Ni was found mainly in the sewage sludge of composting. The order of different fractions of Pb in different raw materials was as follows: F5 (97%) > F1 (2%) > F2 (1%); F5 (99%) > F1 (1%) > F2 (0.1%); and F5 (98%) > F1 (2%) in water hyacinth, sawdust, and cattle manure, respectively. Similar to Ni, the F5 fraction of Pb was also found mainly in all three raw materials. The F3 and F4 fractions of Pb were not detected in water hyacinth and sawdust, whereas F2, F3, and F4 fractions were not detected in cattle manure. The order of different fractions of Cd in different raw materials was as follows: F5 (97%) > F1 (3%) > F2 (0.2%); F5 (99%) > F1 (1%); and F5 (92%) > F1 (8%) in the water hyacinth, sawdust, and cattle manure, respectively. The F5 fraction of Cd was found to be dominant in all the three raw materials. The order of different fractions of Cr in different raw materials was as follows: F5 (83%) > F1 (8%) > F4 (5%) > F2 (4%) > F3 (1%); F5 (59%) > F1 (32%) > F4 (5%) > F2 (3%) > F3 (1%); and F1 (34%) > F5 (31%) > F4 (28%) > F2 (4%) > F3 (2%) in the water hyacinth, sawdust, and cattle manure, respectively. The F5 fraction of Cr was found to be dominant in water hyacinth and sawdust, whereas the F1 fraction of Cr was found mainly in the cattle manure. The total concentration of Cr was found to be very low in cattle manure, whereas its F1 fraction was higher as compared to sawdust and water hyacinth. The detail distribution of Zn, Cu, Mn, Ni, Pb, Cd, and Cr in compost prepared from the different materials is given in the following section.

TABLE 6.1
Chemical Speciation of Heavy Metals (Cd, Cu, Co, and Cr) in Different Wastes Used for Composting Process

Metals	Metal Forms	Raw Materials Taken for Composting	References
Cd	Fe–Mn oxide > carbonate > organic matter > exchangeable > residual	Sewage sludge	Liu et al. (2007)
	Fe–Mn oxide (67.4%) > organic matter (18.4%) > residual (7.3%) > exchangeable (4.17%) > carbonate (2.76%)	Sewage sludge	Gao et al. (2005)
	Residual > oxidizable > reducible > exchangeable	Sewage sludge	Sprynskyy et al. (2007)
	Reducible > residual > exchangeable = carbonate > organic matter	Sewage sludge	Zorpas et al. (2000)
	Carbonate > organically bound > sulfide > exchangeable = adsorbed	Municipal solid waste	Ciba et al. (1999)
	Residual (96.9%) > exchangeable (3%) > carbonate (0.2%)	Water hyacinth	Singh and Kalamdhad (2012)
Cu	Organic matter > residual > exchangeable > carbonate > Fe–Mn oxide	Sewage sludge	Liu et al. (2007)
	Organic matter (51.6%) > residual (43.7%) > carbonate (1.83%) > exchangeable (1.53%) > Fe–Mn oxide (1.29%)	Sewage sludge	Gao et al. (2005)
	Residual (54%) > organic (32.06%) > reducible (9.94%) > carbonate (3.50%) > exchangeable (0.5%)	Sewage sludge	Zorpas et al. (2008)
	Residual > oxidizable > reducible > exchangeable	Sewage sludge	Sprynskyy et al. (2007)
	Oxidizable > residual > reducible > exchangeable > acid extractable	Sewage sludge	Wong and Selvam (2006)
	Residual > organic matter > exchangeable > carbonate > reducible	Sewage sludge	Zorpas et al. (2000)
	Organic matter > residual > carbonate = exchangeable = Fe–Mn oxide	Sewage sludge	Wang et al. (2008)
	Sulfide > carbonate > adsorbed > organically bound > exchangeable > residual > soluble	Municipal solid waste	Ciba et al. (1999)
	Organically bound > organically complexed > solid particles > residual > water soluble > exchangeable	Swine manure	Hsu and Lo (2001)
	Residual (55.6%) > organic matter (34%) > exchangeable (4.7%) > carbonate (3.2%) > reducible (2.4%)	Water hyacinth	Singh and Kalamdhad (2012)

(Continued)

TABLE 6.1 (*Continued*)
Chemical Speciation of Heavy Metals (Cd, Cu, Co, and Cr) in Different Wastes Used for Composting Process

Metals	Metal Forms	Raw Materials Taken for Composting	References
Co	Carbonate = organically bound = sulfide	Municipal solid waste	Ciba et al. (1999)
Cr	Organic matter > residual > exchangeable = Fe–Mn oxide > carbonate	Sewage sludge	Liu et al. (2007)
	Residual (50.9%) > organic matter (42.9%) > exchangeable (3.4%) > carbonate (2%) > Fe–Mn oxide (0.8%)	Sewage sludge	Zheng et al. (2007)
	Organic (47.74%) > residual (32.75%) > reducible (18.08%) > carbonate (1.01%) > exchangeable (0.42%)	Sewage sludge	Zorpas et al. (2008)
	Residual > oxidizable > exchangeable > reducible	Sewage sludge	Sprynskyy et al. (2007)
	Organic matter > residual > reducible > carbonate	Sewage sludge	Zorpas et al. (2000)
	Organic matter > residual > exchangeable > carbonate Fe–Mn oxide	Tannery waste	Shukla et al. (2009)
	Residual (82.7%) > exchangeable (7.9%) > organic matter (4.8%) > carbonate (3.8%) > reducible (0.9%)	Water hyacinth	Singh and Kalamdhad (2012)

TABLE 6.2
Chemical Speciation of Heavy Metals (Ni, Pb, Zn, Fe, and Mn) in Different Wastes Used for the Composting Process

Metals	Metal Forms	Raw Materials Taken for Composting	References
Ni	Residual > exchangeable > organic matter > carbonate > Fe–Mn oxide	Sewage sludge	Liu et al. (2007)
	Residual (32.1%) > organic matter (25.9%) > carbonate (15.3%) > exchangeable (15.1%) > Fe–Mn oxide (11.6%)	Sewage sludge	Zheng et al. (2007)
	Residual (52%) > reducible (22%) > organic (26%) > carbonate (<0.02%) > exchangeable (<0.01%)	Sewage sludge	Zorpas et al. (2008)
	Residual > oxidizable > exchangeable > reducible	Sewage sludge	Sprynskyy et al. (2007)
	Residual > reducible > oxidizable > acid extractable > exchangeable	Sewage sludge	Wong and Selvam (2006)
	Residual > reducible > organic matter > carbonate > exchangeable	Sewage sludge	Zorpas et al. (2000)
	Residual (69.3%) > organic matter (24.0%) > carbonate (3.4%) > exchangeable (2.2%) > Fe–Mn oxide (1%)	Sewage sludge	Wang et al. (2008)
	Carbonate > adsorbed > exchangeable > sulfide > organically bound	Municipal solid waste	Ciba et al. (1999)
	Residual (92.8%) > exchangeable (6.4%) > carbonate (0.8%)	Water hyacinth	Singh and Kalamdhad (2012)
Pb	Organic matter > residual > exchangeable > carbonate > Fe–Mn oxide	Sewage sludge	Liu et al. (2007)
	Residual (75.55%) > reducible (15.58%) > organic (6.87%) > carbonate (0.00%) > exchangeable (0.00%)	Sewage sludge	Zorpas et al. (2008)
	Residual > oxidizable > acid extractable > reducible	Sewage sludge	Wong and Selvam (2006)
	Residual > reducible > carbonate = exchangeable > organic matter	Sewage sludge	Zorpas et al. (2000)
	Carbonate > organically bound > sulfide > adsorbed > exchangeable	Municipal solid waste	Ciba et al. (1999)
	Residual > organic matter > Fe–Mn oxide > carbonate > soluble exchangeable	Sewage sludge	Liu et al. (2009)
	Residual (96.8%) > exchangeable (2.4%) > carbonate (0.8%)	Water hyacinth	Singh and Kalamdhad (2012)

(Continued)

TABLE 6.2 (*Continued*)
Chemical Speciation of Heavy Metals (Ni, Pb, Zn, Fe, and Mn) in Different Wastes Used for the Composting Process

Metals	Metal Forms	Raw Materials Taken for Composting	References
Mn	Reducible (45.15%) > residual (35%) > carbonate (10%) > organic (8.13%) > exchangeable (1.72%)	Sewage sludge	Zorpas et al. (2008)
	Reducible > residual > exchangeable > acid extractable > oxidizable	Sewage sludge	Wong and Selvam(2006)
	Organic matter > residual > carbonate > reducible > exchangeable	Sewage sludge	Zorpas et al. (2000)
	Carbonate > sulfide > organically bound > residual > exchangeable > soluble > adsorbed	Municipal solid waste	Ciba et al. (1999)
	Solid particles > organically complexed > residual > organically bound > exchangeable	Swine manure	Hsu and Lo (2001)
	Residual (40.3%) > exchangeable (27.6%) > organic matter (12.6%) > reducible (11.3%) > carbonate (8.2%)	Water hyacinth	Singh and Kalamdhad (2012)
Zn	Fe–Mn oxide > residual > organic matter > exchangeable	Sewage sludge	Cai et al. (2007)
	Fe–Mn oxide > organic matter > carbonate > residual > exchangeable	Sewage sludge	Liu et al. (2007)
	Organic matter (37.3%) > Fe–Mn oxide (31.0%) > residual (30.6%) carbonate (0.71%) > exchangeable (0.29%)	Sewage sludge	Gao et al. (2005)
	Residual (29.1%) > reducible (26.03%) > organic (23.96%) > carbonate (20%) > exchangeable (1%)	Sewage sludge	Zorpas et al. (2008)
	Oxidizable > residual > reducible > acid extractable > exchangeable	Sewage sludge	Wong and Selvam (2006)
	Residual (38.4%) > organic matter (26.8%) > Fe–Mn oxide (17.9%) > carbonate (13.3%) > exchangeable (3.51%)	Sewage sludge	Wang et al. (2008)
	Carbonate > adsorbed > exchangeable > sulfide > organically bound	Municipal solid waste	Ciba et al. (1999)
	Organically complexed > solid particles > organically bound > exchangeable	Swine manure	Hsu and Lo (2001)
	Residual (63.6%) > organic matter (12.4%) > exchangeable (10.4%) > reducible (8.1%) > carbonate F2 (5.6%)	Water hyacinth	Singh and Kalamdhad (2012)
Fe	Residual (64.19%) > reducible (27.21%) > organic (7.95%) > carbonate (0.38%) exchangeable (0.27%)	Sewage sludge	Zorpas et al. (2008)
	Residual > reducible > organic matter > carbonate > exchangeable	Sewage sludge	Zorpas et al. (2000)
	Organic matter (54.8%) > reducible (30.2%) > residual (13.9%) > exchangeable (0.8%) > carbonate (0.3%)	Water hyacinth	Singh and Kalamdhad (2012)

6.1.1 Zinc (Zn)

The high concentration of Zn was present in bioavailable form increase in the sewage sludge compost that is used to acidic soil (Smith, 2009). A Zn^{2+} is expected to stay in ionic form in a solution it is happened on the basis of its redox potential, it's the redox reaction with other metal ions (Qiao and Ho, 1997). Qiao and Ho (1997) reported that the F2, F3, and F4 fractions of Zn could be attributed to the adsorption–equilibrium interactions in the sludge. The F4 and F5 fractions (stable fractions 0) of Zn were converted into unstable F1 and F2 fractions due to the high rate of oxidation of organic matter in the composting (Chiang et al., 2007; Smith, 2009).

Qiao and Ho (1997) reported that the addition of red mud in a composting mixture can convert the exchangeable fraction of Zn into carbonate fraction by increasing the pH and inhibiting the changes of Zn speciation. The exchangeable fraction of Zn isprecipitated in the form ofzinc hydroxide and zinc carbonate by addition of red mud to the composting mixture through the increasing pH from 5 to 7.2 of the composting biomass. Zn hydroxide is generally dissolves under pH 7.2, so the exchangeable fraction was successfully controlled by pH during the composting of sludge. Cai et al. (2007) stated that the residual fraction of Zn was converted into the reducible and oxidizable fractions during the sewage sludge composting. Venkateswaran et al. (2007) reported that the Zn was mainly present in the reducible fraction followed by the carbonate, oxidizable and residual fractions. High content of Zn in the reducible fraction can be ascribed to diffusion mechanism.

Nomeda et al. (2008) reported that the bioavailable fractions of Zn were enhanced at the maturity of composting process due to the formation of humic substances after oxidation of organic fraction of composting mixture. According to the study done by Gupta and Sinha (2007), Zn was mainly found in the reducible fraction at the end of the composting process of tannery sludge. Ciba et al. (1997) also reported that the carbonate fraction of Zn was increased during the composting process. However, an addition of zinc sulfide to the composting biomass reduced F2 fraction. Around 70% of Zn was bound with residual fraction and oxidizable fraction which will not uptake by plants in any environmental condition (Ciba et al., 1997). Ciba et al. (1999) stated that the organically bound fraction of Zn was found to be dominant in the composting of municipal solid waste (MSW).

According to He et al. (2009b), the exchangeable fraction of Zn was found to be 40–80 times higher at the end of composting of sewage sludge as compared with pig manure compost. Subsequently, the bioavailability factor (BF) of Zn was found to be higher in the composting of sewage sludge as compared with the composting of pig manure. This study also reported that the dominant fractions (organically bound and residual fractions) of Zn were converted into highly mobile fractions (exchangeable and carbonate fractions) during the composting of sewage sludge, whereas in the composting of pig manure most of the stable fractions of Zn were changed into reducible fractions that have less mobility as compared with the exchangeable and carbonate fractions.

A distribution of heavy metals (Cd, Cr, Cu, Zn, Mn, Fe, Ni, Pb and Zn) in the agitated pile and rotary drum composting of water hyacinth is given in Tables 6.3 and 6.4. Singh and Kalamdhad (2012) studied the chemical speciation of Zn during the

TABLE 6.3
Speciation of Zn, Cu, Mn, and Fe in the Agitated Pile and Rotary Drum Composting of Water Hyacinth

Composting Methods	Days	Zn (mg/kg dry matter) F1	F2	F3	F4	F5	Cu (mg/kg dry matter) F1	F2	F3	F4	F5
Agitated pile	0	11.0	25.4	41.7	41.0	39	2.0	2.8	1.4	11.7	10.5
	30	5.7	23.2	46.1	56.8	143	1.6	2.6	1.2	22.8	72.3
Rotary drum	0	4.3	12.6	14.3	21.3	89.7	3.4	1.9	2.0	26.5	31.3
	20	4.0	13.8	26.7	21.2	83.0	0.9	2.1	1.7	15.9	65.1
		Mn (mg/kg dry matter)					**Fe (g/kg dry matter)**				
Agitated pile	0	95	37.8	176.0	27.1	179.5	0.03	0.07	1.6	4.5	1.4
	30	69.5	120.7	220.5	84.2	548.8	0.045	0.028	2.0	6.2	4.2
Rotary drum	0	145.8	89.6	61.4	80.6	142.3	44.7	28.2	2484.5	6386.0	1953.0
	20	82.3	144.8	107.2	59.2	237.9	40.6	33.6	2901.5	4639.2	6058.4

TABLE 6.4
Speciation of Cr, Ni, Cd, and Pb in the Agitated Pile and Rotary Drum Composting of Water Hyacinth

Composting Methods	Days	Cr (mg/kg dry matter) F1	F2	F3	F4	F5	Ni (mg/kg dry matter) F1	F2	F3	F4	F5
Agitated pile	0	29.5	23.1	9.1	11.7	180	16.0	6.6	ND	ND	163.8
	30	22.9	9.8	6.4	7.4	205	11.6	3.4	ND	ND	227.0
Rotary drum	0	5.8	1.8	3.4	15.4	28.5	5.3	3.9	ND	ND	214.6
	20	3.9	1.4	0.8	4.32	62.8	4.0	3.0	ND	ND	227.0
		Cd (mg/kg dry matter)					**Pb (mg/kg dry matter)**				
Agitated pile	0	3.6	2.6	ND	ND	38.3	33.7	13.7	ND	ND	800
	30	2.6	1.5	ND	ND	72.3	25.1	10.0	ND	ND	1375
Rotary drum	0	1.1	1.1	ND	ND	47.5	25.0	15.6	ND	ND	895.4
	20	0.4	0.7	ND	ND	54.1	16.5	8.9	ND	ND	992.5

ND, not detected.

agitated pile composting of water hyacinth. All the mobile fractions (exchangeable, carbonate, reducible, and organically bound fraction) were decreased at the end of the composting process, whereas the stable fraction (residual fraction) was decreased significantly in the thermophilic phase of composting, whereas this fraction was increased at the end of the composting process. Singh and Kalamdhad (2013) studied the chemical speciation of heavy metals during the rotary drum composting of water hyacinth. This study also reported that the F4 and F5 fractions of Zn were decreased, whereas F2 and F3 fractions of Zn were increased (% of total fraction) during the composting process. The F5 fraction of Zn was found to be predominant in both the agitated pile and rotary drum composting of water hyacinth (Singh and Kalamdhad, 2012, 2013). The BF of Zn was decreased from 0.75 (initial) to 0.48 (final) during the agitated pile composting, whereas the BF of Zn was reduced from 0.37 to 0.23 during the rotary drum composting of water hyacinth (Singh and Kalamdhad, 2012, 2013). A decrease in BF of Zn can be ascribed as, bioavailable fractions of Zn were precipitated as zinc hydroxides, zinc carbonates, zinc phosphates, zinc sulfides and zinc organic complexes. Kumpiene et al. (2008) suggested that cation exchange and complexation with organic ligands are the main mechanism involved in controlling mobility of Zn. Cai et al. (2007) also suggested the formation of Zn complex with humic substances at the maturity of composting. Meng et al. (2017) reported that the reducible fraction of Zn found mainly in more than half of the total concentration of Zn in the pig manure composting. The exchangeable and carbonate fractions of Zn decreased slowly during the composting period, whereas the remaining three fractions were increased gradually during the composting period. The exchangeable and carbonate fractions of Zn were converted into more stable fractions in the composting of pig manure. The reducible fraction and organic-matter-bound fraction of Zn were increased from 432.98 to 976.68 mg/kg and from 73.60 to 239.99 mg/kg, respectively, at the end of the composting process, whereas the reducible fraction and organic-matter-bound fraction of Zn were found to be approximately 68% and 17% of the total Zn concentrations in the final compost. From this study, it can be concluded that the potentially available fractions of Zn were decreased suggestively in the composting process. Wu et al. (2017) reported that F1 fractions of Zn were increased up to 1.21–2.05 during the composting of pig manure as compared with the initial composting mixture. The increase in the F1 fraction during this study is due to the decrease in pH (slightly acidic condition). Fractions of F2, F3, and F4 were significantly increased during the composting.

6.1.2 Copper (Cu)

It has been suggested that the oxidizable fraction of Cu was found mainly in the decomposition of sludge, decomposed organic matter or compost is one of the most favorable material for binding of Cu (Yuan et al., 2011). Qiao and Ho (1997) reported that the organic-bound fraction of Cu was found to be dominant at approximately 80% in the composting of sludge. This can be explained by the fact that Cu forms very stable complexes with organic ligands at the end of the composting process. Additionally, the ionic form of Cu could directly form bonds with organic functional groups primarily carboxylic, carbonyl, and phenolic due to immobilization of Cu at the end of the composting process (Qiao and Ho, 1997). Nomeda et al. (2008) reported that in the composting process mobilization of Cu might be due to its high attraction to organic matter. However, the mobile fractions of Cu were suggestively converted into the most stable fraction (residual fraction) in the composting process (Nomeda et al., 2008). Gupta and Sinha (2007) studied speciation of Cu in tannery sludge and reported that Cu was found mainly in the organically bound fraction and residual fraction.

Hsu and Lo (2001) reported that the highest concentration of Cu occurred in the organic-matter-bound fraction due to the formation of the Cu organic matter complex. The exchangeable and carbonate fractions of Cu were decreased slightly at the end of the composting process. This study also reported that around 70% of total Cu was found in the oxidized and organically bound fraction (Hsu and Lo, 2001). Fuentes et al. (2004) studied the distribution of heavy metals in different types of sludges (aerobic, anaerobic, unstabilized, and sludge from a waste stabilization pond) and reported that the Cu was predominantly found in the organically bound fraction. This study also reported that the highest concentration of Cu was found in the organically bound fraction in all the sludges. The organically bound and residual fractions contributed approximately 95% of the total Cu fractions in all selected sludges. From this study, it can be concluded that Cu was connected with tough organic ligands and possibly sealed in quartz, feldsparsetc. (Fuentes et al., 2004). Smith (2009) reported that the mobile fraction of Cu was decreased with an increase in the humic substances and a decrease in the pH value from 7.5 to 6.7 during the composting process

He et al. (2009b) studied the distribution of Cu in the composting of pig manure and reported that eagerly extractible fractions of Cu (exchangeable and carbonate fractions) were reduced significantly, whereas the carbonate fraction of Cu was increased. The reducible and oxidizable fractions of Cu were decreased abruptly during the thermophilic stage of the composting of sewage sludge and remained constant until the maturity of the composting process. However, these fractions (exchangeable and carbonate fractions) were reduced more gradually in the composting of pig manure. The residual fraction of Cu was decreased in both sewage sludge and pig manure composting. This study also reported that oxidizable and residual fractions contributed approximately 50% and 30% of total Cu, respectively, in both composting processes.

Singh and Kalamdhad (2012) reported that the F1, F2, F3, and F4 fractions of Cu were decreased, whereas its F5 fraction was increased (percentage of total fraction) during the agitated pile composting of water hyacinth. The mobile fractions of Cu (F1, F2, F3, and F4 fractions) were converted into the F5 fraction at the end of the composting process. The F1 and F2 fractions of Cu were reduced from 7% and 10% to 1.6% and 2.5%, respectively. Singh and Kalamdhad (2013) stated that the F1, F2, F3, and F4 fractions of Cu were decreased from 5.2%, 2.9%, 3.1%, and 40.8% to 1.1%, 2.6%, 1.9%, and 18.6% of the total fractions of Cu, whereas the F5 fraction was increased from 48.9% to 76% of the total fraction during the composting of water hyacinth in a rotary drum reactor. The mobile fractions of Cu (F1, F2, F3, and F4 fractions) were converted into the F5 fraction at the end of composting process. The reduction in all mobile fractions of Cu in the composting process can be attributed as Cu formed complex with organic functional groups mainly carboxylic, carbonyl and phenolic, resulting reduction in most bioavailable fractions (Qiao and Ho, 1997). The BF of Cu was decreased from 0.63 (initial) to 0.28 (final) during the agitated pile composting, whereas the BF of Cu was reduced from 0.52 to 0.22 during the composting of water hyacinth in a rotary drum. The maximum reduction of the BF of Cu was observed due to presence of hydroxyl and carboxylic groups supplied by cattle manure increased at the binding sites and combined with Cu to form insoluble and immobile complexes (Guan et al., 2011). Furthermore, addition of an appropriate proportion of cattle manure enhanced the composting process and consequently improved the formation of humic substances during the process. Meng et al. (2017) studied speciation of Cu during the composting of pig manure and reported that the organic fraction of Cu was found to be approximately 40%–73% higher than the total fractions of Cu. The exchangeable and carbonate fractions of Cu reduced suddenly in the initial stage of the composting process and then became constant after 28 days, whereas the organic-matter-bound fraction and the residual fraction of Cu were increased significantly from 224.65 to 625.77 and from 17.17 to 101.61 mg/kg, respectively, at the end of composting process. The reducible bound fractions of Cu were changed abruptly during the composting process. This study finally concluded that the potentially available fractions of Cu were decreased suggestively in the composting process. Wu et al. (2017) reported that the residual fraction of Cu was found to be dominant in the final composting of pig manure. An increase in the organic matter degradation and a decrease in the pH could be accountable for the rearrangement of Cu into the residual fraction. This study also reported that the F1 fraction of Cu was decreased, whereas the F4 fraction of Cu was increased suggestively in the final composting. The F2 and F3 fractions of Cu were not observed during the composting process. Lu et al. (2017) studied speciation of Cu and degradation of tetracycline during the composting of water hyacinth with different amendments. This study also reported that speciation of Cu was influenced by the presence of tetracycline due to comparatively low concentration and rapid degradation of tetracycline during the composting process. According to this study, the concentration of easily-available (exchangeable and reducible) fractions of Cu frequently reduced during the composting process. The order of different fractions of Cu in the initial composting of water hyacinth was as follows: residual > oxidizable > reducible > exchangeable. A similar trend of Cu speciation was also observed at the end of composting process. However, the percentage of the exchangeable fraction was decreased and the residual fraction was increased, which specifies a decrease of the bioavailable fractions with an increase of the residual fraction after composting. The exchangeable and reducible fractions of Cu were increased with the increase of Cu in the composting mixture, whereas the residual fraction of Cu was decreased.

6.1.3 Manganese (Mn)

Singh and Kalamdhad (2012) found that the exchangeable and reducible fractions of Mn were decreased in the water hyacinth composting of cattle manure and sawdust, whereas the carbonate, oxidizable, and residual fractions of Mn were increased in the final composting process. Singh and Kalamdhad (2013) reported that the exchangeable and oxidizable fractions of Mn were reduced at the maturity of composting, whereas the carbonate, reducible, and residual fractions were increased. An increase in pH at the maturity of composting can lead to a decrease in the mobility of Mn through precipitation and adsorption (Achiba et al., 2009). The BF of Mn was decreased from 0.65 (initial) to 0.5 during the agitated pile composting (Singh and Kalamdhad, 2012), whereas the BF of Mn was decreased from 0.7 (initial) to 0.6 during the rotary drum composting of water hyacinth (Singh and Kalamdhad, 2013).

Wong and Selvam (2006) reported that the exchangeable, carbonate, and reducible fractions of Mn were increased in control, whereas these fractions of Mn were reduced slightly with the lime addition in the composting of sewage sludge. The organically bound fraction of Mn was not occur much different in the final compost of control and lime treatment. This study also reported that the exchangeable fractions were increased significantly in all treatments; therefore, this study concluded that addition of lime was unable to check the mobility of Mn. Gupta and Sinha (2007) studied the fractionation of Mn in the composting of tannery sludge and described that the concentration of Mn was found mainly in the reducible fraction. Venkateswaran et al. (2007) also reported that the F1 and F2 fractions of Mn contributed the highest proportions as compared to the other fractions (F3, F4 and F5).

6.1.4 Iron (Fe)

Singh and Kalamdhad (2012) reported that the F1, F2, F3, and F4 fractions of Fe were decreased from 0.37%, 0.9%, 21.6%, and 59.3% to 0.36%, 0.2%, 16%, and 50%, respectively, whereas its F5 fraction was increased from 18% to 33.4% of the total fraction of Fe during the composting process. Singh and Kalamdhad (2013) reported that the F1, F2, F3, and F4 fractions of Fe were decreased, whereas its F5 fraction was increased during the composting process. The concentration of Fe was found mainly in the F4 fraction in the final composting of water hyacinth (Singh and Kalamdhad, 2012, 2013). The exchangeable and carbonate fractions of Fe were found to be <2% of total Fe, although the total concentration of Fe was very high. The BF of Fe was decreased from 0.82 (initial) to 0.67 during the agitated pile composting of water hyacinth (Singh and Kalamdhad, 2012), whereas the BF of Fe was decreased from 0.82 (initial) to 0.56 during the rotary drum composting of water hyacinth. The reduction in the BF of Fe could be explained by the fact that humic substances formed a complex compound with Fe (Cai et al., 2007).

6.1.5 Nickel (Ni)

The Ni ions have potential to form complexes with organic ligands just less than Cu^{2+} in the transition metal cations. Therefore, Ni mainly distributed in the oxidized and residual fractions in the sludge composting (Qiao and Ho, 1997). Zheng et al. (2007) reported that the Ni was mainly present in the organically bound fraction in the sewage sludge composting followed by the carbonate and reducible fractions. This study also reported that the highest concentration of Ni was found in the residual fraction. The immobilization of Ni suggested the possible reduction of the risks of the heavy metals during the composting of sewage sludge (Zheng et al., 2007). According to Smith (2009), mobility of Ni was increased with decreasing pH and organic matter content of composting mixture (Smith, 2009). Gupta and Sinha (2007) indicated that the residual fraction of Ni was increased significantly due to alkaline stabilization process in the tannery sludge during the composting process. Amir et al. (2005) reported that the oxidizable fraction of Ni was increased significantly in the composting of sewage sludge. The mobile fractions (F1, F2, F3 and F4 fractions) of Ni decreased, whereas the residual fraction of Ni increased during the composting process (Wong and Selvam, 2006; Zheng et al., 2007; Wang et al., 2008). Smith (2009) stated that the residual fraction of Ni was found to be dominant in the composting of MSW; it contributed up to 50% which represent low accessibility to plants. Singh and Kalamdhad (2012, 2013) reported that the exchangeable and carbonate fractions of Ni were reduced, whereas its residual fraction increased in the composting of water hyacinth with cattle manure. Su and Wong (2003) reported that the residual fraction of Ni was found to be approximately 52% of the total fraction of Ni, whereas the reducible fraction of Ni was found to be the second most dominant fraction in the composting of sewage sludge. Singh and Kalamdhad (2012) reported that the residual fraction contributed approximately 95% in the water hyacinth composting. Both the F1 and F2 fractions contributed <10% of the total fraction of Ni in the final composting of water hyacinth (Singh and Kalamdhad, 2012, 2013). According to Venkateswaran et al. (2007), the exchangeable and carbonate fractions of approximately 15%–30% can cause toxicity to the environment. Singh and Kalamdhad (2012) reported that the residual fraction of Ni was increased significantly due to alkaline stabilization composting biomass in the final composting of water hyacinth, whereas the reducible and oxidizable fractions of Ni were not observed in the composting of water hyacinth (Singh and Kalamdhad, 2012, 2013). The BF of Ni was decreased from 0.12 (initial) to 0.06 (final) during the agitated pile composting of water hyacinth (Singh and Kalamdhad, 2012), whereas the BF of Ni was reduced from 0.041 (initial) to 0.014 (final) during the rotary drum composting of water hyacinth (Singh and Kalamdhad, 2013). This reduction of the BF of Ni confirmed that the addition of an appropriate amount of cattle manure in the composting of water hyacinth could effectively reduce the availability of Ni for plant uptake (Singh and Kalamdhad, 2012, 2013). Ni can form organometallic complex at the maturity of the composting process (Qiao and Ho, 1997).

6.1.6 Lead (Pb)

According Qiao and Ho (1997), Pb has a high potential to occupy the adsorption sites present in the clay materials. The organically bound and residual fractions of Pb were found to be dominant in the composting of sewage sludge, whereas the exchangeable, carbonate, and reducible fractions of Pb were found uniformly scattered in the final composting (Qiao and Ho, 1997). Ciba et al. (1999) reported that the carbonate and oxidizable fractions of Pb were found mainly in the composting of MSW. Venkateswaran et al. (2007) reported that the oxidizable fraction of Pb was found to be the most dominant fraction as compared to the other fraction and it was released during the composting process. Qiao and Ho (1997) concluded that bioavailability of Pb in the compost of sewage sludge decreased due to binding of Pb fractions with dissolved organic carbon (DOC). Positively charged Pb absorbed on the clay surface and on the surface of dissolved organic carbon in sewage sludge through formation of complex.

He et al. (2009b) reported that contents of five fractions (obtained after SE) of Pb were found to be lower in the compost of swine manure compost as compared to the compost of sewage sludge, however, trend of concentration of all five fraction was observed similar in compost of both sewage sludge and swine manure. The exchangeable, reducible, and residual fractions of Pb were reduced gradually at the end of the composting process, whereas the carbonate-bound fraction of Pb was increased significantly at the maturity of composting. The organically bound fraction of Pb was varied during the thermophilic stage of the composting process, but it was lower in the final composting. Fuentes et al. (2004) reported that the F3 fraction of Pb was present in highest concentration in raw sludge, it contributed around 40%, while in other sludges, the Pb was spreaded in oxidizable and residual fraction, these fractions contributed around 73% during oxidation of the aerobic sludge.

Qiao and Ho (1997) discovered that the Pb stabilized suggestively in the composting mixture at end of composting process. This study also reported that the exchangeable and carbonate fractions of Pb were transformed into the oxidizable fraction in the 50 days' period of the composting process. Wong and Selvam (2006) reported that three mobile fractions (the carbonate, reducible, and oxidizable fractions) of Pb were increased, whereas the residual fraction of Pb was decreased at the maturity of the composting. The exchangeable fraction of Pb was not found at the end of composting process. The F1, F2, F3 and F4 fractions of Pb were not changed significantly after lime addition to the composting biomass, whereas the residual fraction of Pb was increased at the maturity of composting process. Singh and Kalamdhad (2012) reported that the F1 and F2 fractions of Pb were reduced, whereas its F5 fraction was increased in the final composting of water hyacinth. Amir et al. (2005) also reported that the mobile fractions of Pb were reduced during the composting of sewage sludge. Singh and Kalamdhad (2013) reported that the F1 and F2 fractions of Pb were decreased, whereas its F5 fraction was increased (percentage of total fraction) during the rotary drum composting of water hyacinth.

The F1 and F2 fractions of Pb were contributed <5% of the total fraction of Pb, whereas the residual fraction of Pb was found mainly in the final composting of water hyacinth. The F3 and F4 fractions of Pb were not found in the final composting of water hyacinth (Singh and Kalamdhad, 2012, 2013). The mobility of Pb was decreased during the composting process of water hyacinth (Singh and Kalamdhad, 2013). A similar trend of Pb mobility during the composting of sewage sludge was also reported by Wong and Selvam (2006). The BF of Pb was reduced from 0.06 (initial) to 0.03 (final) during the agitated pile composting of water hyacinth, whereas the BF of Pb was decreased from 0.043 (initial) to 0.025 (final) during the rotary drum composting of water hyacinth (Singh and Kalamdhad, 2012).

6.1.7 Cadmium (Cd)

Cd is mainly found in the moderately carbonate fraction and is potentially available to plants (Ciba et al., 1999). Fuentes et al. (2004) reported that the Cd is mainly found in the organic-matter-bound fraction in waste stabilization pond and anaerobic sludge, whereas the exchangeable fraction of Cd contributed around 15% of the total Cd in the anaerobic sludge. Comparatively high proportions of easily-available fractions of Cd were attained in sludge which was least stabilized (Fuentes et al., 2004). The F5 fraction of Cd was increased, whereas the F1 and F2 fractions were reduced in the final composting of water hyacinth (Singh and Kalamdhad, 2012, 2013). The F1 and F2 fractions of Cd were decreased, however F5 fraction was increased in the final compost(Haroun et al., 2007). Hanc et al. (2009) reported that F1 and F2 fractions of Cd were found mainly in the composting of sewage sludge. The BF of Cd was reduced from 0.063 (initial) to 0.053 (final) during the agitated pile composting of water hyacinth (Singh and Kalamdhad, 2012). However, the BF of Cd was reduced from 0.045 (initial) to 0.019 (final) during the rotary drum composting of water hyacinth (Singh and Kalamdhad, 2013). Reduction in the BF of Cd in the final compost of water hyacinth due to binding of bioavailable fractions of Cd to carboxylic and phenolic functional groups of final compost, therefore Cd ions were restrained in an inflexible inner-sphere composite (Qiao and Ho, 1997).

6.1.8 Chromium (Cr)

Cr is the tenth rich metal of the earth and is found in nature as either Cr (III) or Cr (VI) (Zhou et al., 2006). Cr (VI) is a dominant oxidizing agent and is highly toxic to the environment. Cr (III) is approximately 10–100 times less toxic as compared with Cr (VI), although Cr (III) is an essential micronutrient and is considered a hazardous species (Zhou et al., 2006). Due to the low solubility, mobility, and bioavailability of Cr (III), this species has less toxic or no toxic effects on mammalian, aquatic environment, and plants (Zhou et al., 2006). The Cr (III) exhibited a reduction in solubility/extractability in the organic material stabilization process followed by increasing level of humification in the composting of MSW (Ciavatta et al., 1993). Zhou et al. (2006) stated that the removal of Cr from tannery sludge is important to use this sludge or its compost as soil amendment, because organic ligands and/or acid conditions may enhance mobility of Cr (III) in soils. The presence of MnO in soil oxidized it into the more toxic form of Cr (VI) (Zhou et al., 2006).

Smith (2009) stated that in the composting of MSW, Cr was mainly present in the residual fraction in the highly stable form. Consequently, it has enormously low solubility and bioavailability. Shukla et al. (2009) reported that the residual fraction of Cr was found to be higher in the initial composting mixture, whereas this fraction was converted into the carbonate fraction at the maturity of composting of tannery treated biomass. Gupta and Sinha (2007) reported that the reducible fraction of Cr was found mainly in tannery sludge. Zheng et al. (2007) studied speciation of Cr in the composting of sewage sludge and reported that the four mobile fractions (the F1, F2, F3 and F4 fractions) of Cr were reduced at the end of composting process, whereas the residual fraction of Cr was increased, results represented that four mobile fractions (F1, F2, F3 and F4 fractions) of Cr were converted into the residual fraction in composting process. Transformation of different mobile fractions of Cr into less mobile fractions in sewage sludge suggest reduction of potential risks of the Cr in the compost (Zheng et al., 2007). Qiao and Ho (1997) reported that the reducible and carbonate fractions of Cr were transformed into the organically bound fraction in the composting process. Fuentes et al. (2004) reported that the oxidizable fraction of Cr was found mainly in waste stabilization pond and sewage sludge composting, whereas the residual fraction of Cr was found in other sludges. Xu et al. (2012) reported F4 and F5 fractions of Cr contributed in the range of 75%–97% of total Cr, results of this study represented that Cr will not release in the environment under normal environmental conditions when utilization of fly ash-sludge to soil.

Singh and Kalamdhad (2012) reported that the F5 fraction of Cr was increased, whereas the F1, F2, F3, and F4 fractions of Cr were reduced from 11.5%, 9.1%, 3.6%, and 4.6% to 9.1%, 3.7%, 2.4%, and 2.8% of total fractions of Cr during the composting process. Singh and Kalamdhad (2013) reported that all the mobile fractions (exchangeable, carbonate, reducible, and oxidizable) of Cr were reduced, whereas its residual fraction was increased at the end of the composting process. The BF of Cr was reduced from 0.29 to 0.22 in the agitated pile composting of water hyacinth (Singh and Kalamdhad, 2012), whereas the BF of Cr was reduced from 0.48 (initial) to 0.14 (final) during the rotary drum composting of water hyacinth. A decrease in BF of Cr in the final compost of water hyacinth can attributed as F1 and F2 fractions of Cr bound to numerous organic functional groups occurring in the humic substances, however F3 and F4 fractions were transformed into F5 fraction in the composting.

6.2 CONCLUSION

Heavy-metal toxicity depends on the mobile fractions of metals present in the final composting. The chemical speciation of heavy metals allows the evaluation of the bioavailability of the metals on the basis of their bonding strength, either in ionic form or complexed by organic matter. Bioavailability and speciation of heavy metals depend on physicochemical properties organic wastes such as mineralization of organic matter, pH, etc., instead of total metal contents. Addition of cattle manure to the composting mixture enhances the organic matter degradation and humification process; therefore, it decreased the key bioavailable fractions (exchangeable and carbonate) of heavy metals in the composting process of various waste materials. The composting process is highly efficient for the stabilization of waste materials and for the immobilization of heavy metals. Bioavailable fractionations of heavy metals increase during the thermophilic stage of composting, whereas these fractions are converted into less mobile fractions at the maturity of the composting.

REFERENCES

Achiba, W.B., Gabteni, N., Lakhdar, A., Laing, G.D., Verloo, M., Jedidi, N., and Gallali, T. 2009. Effects of 5-year application of municipal solid waste compost on the distribution and mobility of heavy metals in a Tunisian calcareous soil. *Agriculture, Ecosystem and Environment* 130: 156–163.

Amir, S., Hafidi, M., Merlina, G., and Revel, J.C. 2005. Sequential extraction of heavy metals during composting of sewage sludge. *Chemosphere* 59: 801–810.

Cai, Q.Y., Mo, C.H., Wu, Q.T., Zeng, Q.Y., and Katsoyiannis, A. 2007. Concentration and speciation of heavy metals in six different sewage sludge-composts. *Journal of Hazardous Material* 147: 1063–1072.

Chiang, K.Y., Huang, H.J., and Chang, C.N. 2007. Enhancement of heavy metal stabilization by different amendments during sewage sludge composting process. *Journal of Environmental Engineering and Management* 17(4): 249–256.

Ciavatta, C., Govi, M., Simoni, A., and Sequi, P. 1993. Evaluation of heavy metals during stabilization of organic matter in compost produced with municipal solid wastes. *Bioresource Technology* 43: 147–153.

Ciba, J., Korolewicz, T., and Turek, M. 1999. The occurrence of metals in composted municipal wastes and their removal. *Water Air and Soil Pollution* 111: 159–170.

Ciba, J., Zolotajkin, M., and Cebula, J. 1997. Changes of chemical forms of Zinc and Zinc sulfide during the composting process of municipal solid waste. *Water Air and Soil Pollution* 93: 167–173.

Fuentes, A., Llorens, M., Saez, J., Aguilar, M.I., Soler, A., Ortuno, J.F., and Meseguer, V.F. 2004. Simple and sequential extractions of heavy metals from different sewage sludges. *Chemosphere* 54: 1039–1047.

Gao, D., Zheng, G.-D., Chen, T.-B., Luo, W., Gao, W., and Zhang, Y.-A. 2005. Changes of Cu, Zn, and Cd speciation in sewage sludge during composting. *Journal of Environmental Science* 17: 957–961.

Guan, T.X., He, H.B., Zhang, X.D., and Bai, Z. 2011. Cu fractions, mobility and bioavailability in soil-wheat system after Cu-enriched livestock manure applications. *Chemosphere* 82: 215–222.

Gupta, A.K., and Sinha, S. 2007. Phytoextraction capacity of the plants growing on tannery sludge dumping sites. *Bioresource Technology* 98: 1788–1794.

Hanc, A., Tlustos, P., Szakova, J., and Habart, J. 2009. Changes in cadmium mobility during composting and after soil application. *Waste Management* 29: 2282–2288.

Hargreaves, J.C., Adl, M.S., and Warman, P.R. 2008. A review of the use of composted municipal solid waste in agriculture. *Agriculture, Ecosystem and Environment* 123: 1–14.

Haroun, M., Idris, A., and Omar, S.R.S. 2007. A study of heavy metals and their fate in the composting of tannery sludge. *Waste Management* 27: 1541–1550.

He, M., Li, W, Liang, X., Wu, D., and Tian, G. 2009a. Effect of composting process on phytotoxicity and speciation of copper, zinc and lead in sewage sludge and swine manure. *Waste Management* 29: 590–597.

He, M., Tian, G., and Liang, X. 2009b. Phytotoxicity and Speciation of Copper, Zinc and Lead during the Aerobic Composting of Sewage Sludge. *Journal of Hazardous Material* 163: 671–677.

Hsu, J.H., and Lo, S.L. 2001. Effects of composting on characterization and leaching of copper, manganese, and zinc from swine manure. *Environmental Pollution* 114: 119–127.

Kumpiene, J., Lagerkvist, A., and Maurice, C. 2008. Stabilization of As, Cr, Cu, Pb and Zn in soil using amendments -A review. *Waste Management* 28: 215–225.

Khaled, E.M. 2004. Distribution of different fractions of heavy metals in desert sandy soil amended with composted sewage sludge. The International Conference on Water Resources and Arid Environment, Prince Sultan Research Center For Environment, Water and Desert, King Saud University, Riyadh, Kingdom of Saudi Arabia (5–8 December, Abst.45).

Li, X.D., Poon, C.S., Sun, H., Lo, I.M.C., and Kirk, D.W. 2001. Heavy metal speciation and leaching behaviors in cement based solidified/stabilized waste materials. *Journal of Hazardous Materials* 82: 215–230.

Liu, Y., Ma, L., Li, Y., and Zheng, L. 2007. Evolution of heavy metal speciation during the aerobic composting process of sewage sludge. *Chemosphere* 67: 1025–1032.

Liu, J., Xu, X., Huang, D., and Zeng, G. 2009. Transformation behavior of lead fractions during composting of lead-contaminated waste. *Transactions of Nonferrous Metals Society of China* 19: 1377–1382.

Lu, X., Liu, L., Fan, R., Luo, J., Yan, S., Rengel, Z., and Zhang, Z. 2017. Dynamics of copper and tetracyclines during composting of water hyacinth biomass amended with peat or pig manure. *Environmental Science and Pollution Research* 24: 23584–23597.

Meng, J., Wang, L., Zhong, L., Liu, X., Brookes, P.C., Xu, J., and Chen, H. 2017. Contrasting effects of composting and pyrolysis on bioavailability and speciation of Cu and Zn in pig manure. *Chemosphere* 180: 93–99.

Nair, A., Juwarkar, A.A., and Devotta, S. 2008. Study of speciation of metals in an industrial sludge and evaluation of metal chelators for their removal. *Journal of Hazardous Material* 152: 545–553.

Nemeth, T., Bujtas, K., Csillag, J., Partay, G., Lukacs, A., and Van, M.T. 1999. Distribution of ecologically significant fraction of selected heavy metals in the soil profile, in H.M. Selim and I.K. Iskandar (eds.) *Fat and Transport of Heavy Metals in the Vadose Zone*, pp. 251–271. Lewis Publisher CRC Press, Boca Raton, FL.

Nomeda, S., Valdas, P., Chen, S.Y., and Lin, J.G. 2008. Variations of metal distribution in sewage sludge composting. *Waste Management* 28: 1637–1644.

Pare, T., Dinel, H., and Schnitzer, M. 1999. Extractability of trace metals during co-composting of biosolids and municipal solid wastes. *Biology and Fertility of Soils* 29: 31–37.

Qiao, L., and Ho, G. 1997. The effects of clay amendment and composting on metal speciation in digested sludge. *Water Research* 31(5): 951–964.

Shukla, O.P., Rai, U.N., and Dubey, S. 2009. Involvement and interaction of microbial communities in the transformation and stabilization of chromium during the composting of tannery effluent treated biomass of *Vallisneria spiralis L. Bioresource Technology* 100: 2198–2203.

Singh, J., and Kalamdhad, A.S. 2012. Concentration and speciation of heavy metals during water hyacinth composting. *Bioresource Technology* 124: 169–179.

Singh, J., and Kalamdhad, A.S. 2013. Effect of rotary drum on the speciation of heavy metals during water hyacinth composting. *Environmental Engineering Research* 18: 177–189.

Smith, S.R. 2009. A critical review of the bioavailability and impacts of heavy metals in municipal solid waste composts compared to sewage sludge. *Environment International* 35: 142–156.

Sprynskyy, M., Kosobucki, P., Kowalkowski, T., and Buszewsk, B. 2007. Influence of clinoptilolite rock on chemical speciation of selected heavy metals in sewage sludge. *Journal of Hazardous Material* 149: 310–316.

Su, D.C., and Wong, J.W.C. 2003. Chemical speciation and phytoavailability of Zn, Cu, Ni and Cd in soil amended with fly ash-stabilized sewage sludge. *Environmental International* 29: 895–900.

Tessier, A., Campbell, P.G.C., and Bisson, M. 1979. Sequential extraction procedures for the speciation of particulate trace metals. *Analytical Chemistry* 58: 844–851.

Venkateswaran, P., Vellaichamy, S., and Palanivelu, K. 2007. Speciation of heavy metals in electroplating industry sludge and wastewater residue using inductively coupled plasma. *International Journal of Environmental Science and Technology* 4(4): 497–504.

Walter, I., Martinez, F., and Cala, V. 2006. Heavy metal speciation and phytotoxic effects of three representative sewage sludge for agricultural uses. *Environmental Pollution* 139: 507–514.

Wang, X., Chen, L., Xia, S., and Zhao, J. 2008. Changes of Cu, Zn, and Ni chemical speciation in sewage sludge co-composted with sodium sulfide and lime. *Journal of Environmental Science* 20: 156–160.

Wong, J.W.C., and Selvam, A. 2006. Speciation of heavy metals during co-composting of sewage sludge with lime. *Chemosphere* 63: 980–986.

Wu, S.H., Shen, Z.Q., Yang, C.P., Zhou, Y.X., Li, X., Zeng, G.M., Ai, S.J., and He, H.J. 2017. Effects of C/N ratio and bulking agent on speciation of Zn and Cu and enzymatic activity during pig manure composting. *International Biodeterioration and Biodegradation* 119: 429–436.

Xu, J.Q., Yu, R.L., Dong, X.Y., Hu, G.R., Shang, X.S., Wang, Q., and Li, H.W. 2012. Effects of municipal sewage sludge stabilized by fly ash on the growth of manilagrass and transfer of heavy metals. *Journal of Hazardous Materials* 217–218: 58–66.

Yuan, X., Huang, H., Zeng, G., Li, H., Wang, J., Zhou, C., Zhu, H., Pei, X., Liu, Z., and Liu, Z. 2011. Total concentrations and chemical speciation of heavy metals in liquefaction residues of sewage sludge. *Bioresource Technology* 102: 4104–4110.

Zheng, G.D., Chen, T.B., Gao, D., and Luo, W. 2007. Stabilization of nickel and chromium in sewage sludge during aerobic composting. *Journal of Hazardous Materials* 142: 216–221.

Zhou, S.P., Zhou, L.X., Wang, S.M., and Fang, D. 2006. Removal of Cr from tannery sludge by bioleaching method. *Journal of Environmental Science* 18(5): 885–890.

Zorpas, A.A., Constantinides, T., Vlyssides, A.G., Haralambous, I., and Loizidou, M. 2000. Heavy metal uptake by natural zeolite and metals partitioning in sewage sludge compost. *Bioresource Technology* 72: 113–119.

Zorpas, A.A., Vassilis, I., and Loizidou, M. 2008. Heavy metals fractionation before, during and after composting of sewage sludge with natural zeolite. *Waste Management* 28: 2054–2060.

7 Effects of Chemical Amendments on Bioavailability and Fractionation of Heavy Metals in Composting

7.1 EFFECTS OF CHEMICAL AMENDMENTS

Sprynskyy et al. (2007) reported that all four fractions of metals obtained through the sequential procedure were decreased due to the addition of the natural clinoptilolite to the sewage sludge. The mobile fractions of metals were decreased approximately 87% for Cd, 64% for Cr, 35% for Cu, and 24% for Ni due to the addition of 9.09% of the clinoptilolite. The total concentration of metals was also reduced approximately 11%, 15%, 25%, 41%, and 51% for Cu, Ni, Cr, Cd, and Pb, respectively, after addition of 9.09% clinoptilolite to the sludge. The clinoptilolite rock can be considered as a suitable material for the immobilization of heavy metals by bonding with absorbed forms of the metals. The F1 fraction of Cr was also reduced by 49% after the zeolite addition during the composting of sewage sludge (Sprynskyy et al., 2007). Stylianou et al. (2008) stated that the Zn was reduced by approximately 94%, Cu by 60%, Cr by 82%, and Ni by 69%, whereas Mn was reduced by approximately 48%. The F1 fraction of metals may adsorbed on the surface of zeolite, resulting immobilization of the metals during the composting process. According to Stylianou et al. (2008), hydration of the cation is also important during the composting process. The order of metals in clinoptilolite–water ion exchange systems was as follows: $Pb^{2+} > Cr^{3+} > Fe^{3+} > Cu^{2+}$, and $Pb^{2+} > Zn^{2+} > Cu^{2+}$. Zorpas et al. (2008) reported that the percentage of Cr is higher in the oxidizable fraction during the thermophilic phase than the other fractions, whereas it has been transferred into the residual fraction at the end of composting process in thermophilic phase. Zeolite bound all easily available form of metal which are generally considered as mobile forms) at the end of the composting process (Zorpas et al., 2008). Lime is a well-known alkaline material used to decrease the heavy-metal mobility in the composting process (Chiang et al., 2007). The application of lime during the composting process which increased the pH of the composting biomass can be very useful for the minimization of most of the bioavailable fraction of heavy metals (Singh and Kalamdhad, 2013). Addition of lime during the composting process providing a buffering in the composting biomass in contradiction of the reduced in pH during degradation of biomass, and also adding

Ca, which could progress the metabolic activity of microorganisms. Subsequently, temperature of composting biomass and CO_2 evolution are increasing without any undesirable effects on the microbial community (Gabhane et al., 2012). An application of lime in the composting process could highly efficient for the reduction in bioavailability of heavy metals in composting process can be attributed the formation insoluble carbonate salts (Fang and Wong, 1999).

Villasenor et al. (2011) stated that frameworks of zeolite are based on three-dimensional SiO_4 and AlO_4 tetrahedra. The Al^{3+} is found mainly in the midpoint of the tetrahedron with four oxygen atoms, whereas the removal of Si^{4+} by Al^{3+} from the lattice developed a negative charge. The inter-changeable elements, such as Na^+, K^+, or Ca^{2+}, develop a net negative charge. These positive-charge elements are transferable with definite cations in solutions such as Pb, Th, Cd, Zn, Mn, and NH_4^+ (Sprynskyy et al., 2007; Villasenor et al., 2011). Natural zeolite has been widely used to reduce the bioavailability of heavy metals due to its sorption and exchangeable properties towards the heavy metals in the composting of sewage sludge (Zorpas et al., 2000; Sprynskyy et al., 2007). Natural zeolite has potential capacity to uptake of heavy metals are present in easily available fractions, and exchange of sodium and potassium (Zorpas et al., 2000). Sprynskyy et al. (2007) reported that natural zeolites (clinoptilolite) in particular seem to be appropriate amendment materials due to their sorption and exchangeable properties towards the heavy metals in the composting of sewage sludge. The addition of the clinoptilolite rock to sewage sludge might change chemical speciation of heavy metals in composts and decrease their mobility and bioavailability (Sprynskyy et al., 2007). Villasenor et al. (2011) reported that the zeolites retained 100% of the Ni, Cr, and Pb fractions present in the sludge. Stylianou et al. (2008) determined that clinoptilolites displayed the following metal retention selectivity: $Zn^{+2} > Cr^{+3} > Ni^{+2} > Cu^{+2} > Mn^{+2}$. Table 6.1 shows the application of different amendments in the composting of different metals.

7.2 EFFECT OF LIME ON THE BIOAVAILABILITY AND SPECIATION OF HEAVY METALS

7.2.1 Effects of Lime on Water Solubility

Table 7.1 displays the variations in the water-soluble fractions of Cr, Cu, Mn, Fe, and Zn in the agitated pile composting of the horizontal rotary drum reactor. The concentration of water-soluble heavy metals was lower than their total contents in the final composting, but even very low concentrations of water-soluble heavy metals may be toxic to plants (Fang and Wong, 1999). The sequence of water-soluble metals in the composting of water hyacinth was as follows: Cu > Zn > Mn > Cr > Fe. The application of lime caused a great reduction in water extractability of metals (Zn, Cu, Mn, Fe, and Cr) at the end of the agitated pile composting of water hyacinth. The water-solubility fractions of Cr, Zn, Mn, and Fe with lime treatment were decreased in the composting of water hyacinth in the horizontal rotary drum reactor. A significant decrease in water solubility of metals was observed in lime treatment due to the higher degradation of organic matter of composting biomass followed by the development of soluble forms of carbonates and organometallic complexes (Fang and

TABLE 7.1
Effects of Waste Lime on Water Solubility of Heavy Metals during the Agitated Pile Composting and Rotary Drum Composting Water Hyacinth

Composting Methods	Days	Water-Soluble Metals Concentration (mg/kg)							
		Zn	Cu	Mn	Fe	Ni	Pb	Cd	Cr
Agitated pile	0	3.07	2.07	1.8	11.51	ND	ND	ND	3.09
	30	1.04	0.55	2.98	10.7	ND	ND	ND	0.5
Rotary drum	0	3.70	2.97	3.04	19.96	ND	ND	ND	2.21
	20	1.32	1.10	1.13	11.77	ND	ND	ND	0.35

ND, not detected.

Wong, 1999). Cd, Ni, and Pb were not found in water-soluble form in lime treatment (Singh and Kalamdhad, 2013, 2014a). Fang and Wong (1999) reported that water-soluble contents of Zn and Cu were decreased in the compost prepared from sewage sludge. Additionally, Zn was found mainly as Zn^{2+} species at pH <7.5 and is predictable to form zinc hydroxide at pH 7.5–11.5. Furthermore, Zn hydroxy-complexes, a calcium–Zn complex hydrated complex [$CaZn_2$ $(OH)_6.2H_2O$] may be formed at pH ranging from 7 to 12; therefore, the solubility of Zn was decreased due to development of a complex with organic matter (Chen et al., 2009). The lime addition to the composting mixture was very effective in decreasing bioavailability of heavy metals (Zn, Cu, Mn, Fe, and Cr) in both agitated and horizontal rotary drum reactor composting of water hyacinth (Singh and Kalamdhad, 2013, 2014a). The water solubility of metals was decreased due to the formation of less-soluble carbonate in the composting process (Fang and Wong, 1999). In the aqueous solution, lime [calcium hydroxide, $Ca(OH)_2$] breaks down into Ca^{2+} and OH^- (Montes-Hernandez et al., 2009). The metal ion (M^{2+}) combined with OH^- forms metal hydroxides; these hydroxides complexed with metals can be adsorbed on the surface of colloids formed due to the degradation of organic matter, leading to a decrease in the metal solubility in the composting biomass (Garau et al., 2007).

Addition of lime to the composting mixture raised the pH of the mixture and subsequently decreased the bioavailability and eco-toxicity of heavy metals (Baker et al., 2011). Additionally, the water extractability of heavy metals was reduced as microbial biomass adsorbs the available form of metals in the composting biomass or formed metal complex with the humic substances developed at the end of the composting process (Castaldi et al., 2006; Cai et al., 2007; Singh and Kalamdhad, 2013). Singh and Kalamdad (2013, 2014a) reported that water solubility of Ni and Pb was not found in the composting of water hyacinth. Fuentes et al. (2004) reported that the water-soluble fraction of Ni was found to be less than 1 mg/kg in sewage sludge.

7.2.2 Effects of Lime on DTPA Extractability

Table 7.2 displays the DTPA extraction effectiveness of heavy metals during agitated pile and horizontal rotary drum reactor composting of water hyacinth with lime

TABLE 7.2
Effects of Waste Lime on DTPA Extractability of Heavy Metals during the Agitated Pile Composting and Rotary Drum Composting Water Hyacinth

Composting Methods	Days	DTPA Extraction Concentration of Metals (mg/kg)							
		Zn	Cu	Mn	Fe	Ni	Pb	Cd	Cr
Agitated pile	0	44.86	10.6	163.05	222.35	2.65	ND	ND	5.55
	30	34.82	4.72	130.3	226.6	0.71	ND	ND	2.57
Rotary drum	0	41.65	13.9	177.25	381.2	2.05	ND	ND	3.25
	20	18.58	4.9	132	265.75	0.48	ND	ND	1.17

ND, not detected.

(Singh and Kalamdhad, 2013, 2014a). The concentration of Cr, Cu, Mn, Fe, Ni, and Zn contents was decreased in the agitated pile and horizontal rotary drum reactor composting of water hyacinth with lime addition. The concentration of Pb and Cd in DTPA-extractable solution was not observed in the final compost of water hyacinth with lime treatment. The reduction in DTPA extractability of metals is likely due to the development of less-soluble carbonates and hydroxides of metals with lime (Wong and Selvam, 2006). Khan and Jones (2009) described that lime is highly effective to decrease bioavailability of Zn by reducing its mobility in Cu mine tailing soil.

DTPA extractability found the highest for Mn followed by Zn, Cu, Fe, Cr, and Ni in the compost of water hyacinth. DTPA extraction of metals was decreased successfully by the addition of lime in composting of water hyacinth. This is likely due to an increase in pH of composting biomass by lime addition. Fang and Wong (1999) also reported that the concentration of Zn in DTPA-extractable solution was found to be higher in the composting of sewage sludge. The concentration of Mn was increased significantly in the final compost of water hyacinth.

Wong and Selvam (2006) reported that the DTPA extractability of Cu, Mn, Ni, Pb, and Zn was reduced with increasing lime at the end of composting of sewage sludge. The concentrations of Ni and Zn were not changed significantly in both control and lime treatment. The concentration of DTPA-extractable Zn was found to be higher followed by Mn, Cu, Pb, and Ni in the composting of sewage sludge. Chiang et al. (2007) studied the bioavailability of metals in the composting of sewage sludge and reported that the fractions of Pb, Cu, and Zn in DTPA-extracted solution was significantly reduced with increasing pH in the composting process with lime treatment. Fang and Wong (1999) reported that a significant decrease in the initial extractable Ni content was only obtained at the high lime-amendment levels (51%). Fuentes et al. (2004) informed DTPA extraction efficiency of metals was found to be approximately 37% for Ni and Zn and 30% for Cd in the composting of sewage sludge. The maximum extraction of these metals was found approximately 0.75% for Ni, 20.2% for Zn, and 0.0% for Cd in the pile composting of water hyacinth, whereas these metals contributed as 0.40% for Ni, 33% for Zn, and 0% for Cd in the horizontal rotary drum reactor composting of water hyacinth with lime addition (Singh and Kalamdhad, 2013, 2014a).

7.2.3 Effects of Lime of Speciation of Heavy Metals

Table 7.3 shows the speciation of Zn, Cu, Mn, and Fe in lime treatment during agitated pile composting and rotary drum of composting of water hyacinth. Singh and Kalamdhad (2016) reported that all the mobile fractions (the exchangeable, carbonate, reducible, and oxidizable fractions) of Zn were decreased, whereas its residual fraction was increased in the pile composting of water hyacinth with lime addition.

The sequence of different fractions of Zn in the lime-treated compost was as follows: F5 > F3 > F4 > F2 > F1. The F4 fraction was transformed into F3 and F5 fractions in the lime-treated compost. The Zn was mainly present in the residual fraction in the lime-amended compost. These results confirm that the mobile fractions of Zn were converted into the stable fraction due to lime addition. Singh et al. (2015) reportedthat higher decrease of the F2 fraction was observed approximately 78% and the F3 fraction decreased approximately 57% were detected in lime-treated compost of the water hyacinth. The F5 fraction of Zn was enhanced in the lime-amended compost. Significant reduction of the F4 fraction might be due to conversion of this fraction into the F3 and F5 fractions during the composting process. The F5 fraction of Zn was found to be dominant in lime treatments. Zhu et al. (2011) stated that the reducible and oxidizable fractions of Zn were found to be dominant in sewage sludge. The reduction of the F1, F2, F3, and F4 fractions (percentage of total fraction) in lime-treated compost might be due to cation exchange and complexation by decomposed organic matter (Cai et al., 2007; Kumpiene et al., 2008). Most of the mobile fractions of Zn were transformed into the stable fraction (the

TABLE 7.3
Speciation of Zn, Cu, Mn, and Fe in Lime Treatment during Agitated Pile Composting and Rotary Drum of Composting of Water Hyacinth

		Speciation of Heavy Metals									
		Zn (mg/kg)					Cu (mg/kg)				
Composting Methods	Days	F1	F2	F3	F4	F5	F1	F2	F3	F4	F5
Agitated pile	0	7.61	7.56	42.07	24.66	40.28	6.26	3.65	0.65	15.69	23.0
	30	3.02	4.11	54.55	21.62	109.0	1.99	1.14	0.88	12.41	64.8
Rotary drum	0	6.44	16.64	35.1	21.34	50.95	6.03	2.80	1.74	12.45	12.78
	20	2.12	4.15	35.95	9.90	93.5	1.13	1.08	0.31	9.10	47.45
		Mn (mg/kg)					Fe (mg/kg)				
Agitated pile	0	13.0	126.2	194.2	58.7	36.5	27.4	26.9	2788	4477	1655
	30	33.2	97.7	285.1	43.1	189.0	13.1	11.9	1929	3741	13925
Rotary drum	0	41.38	130.9	119.0	19.87	49.2	121.5	23.1	2126	2517	1567
	20	20.60	72.45	173.2	10.94	230.0	148.9	15.3	1208	1737	7433

Note: Composition of treated compost: water hyacinth (90 kg), sawdust (15 kg), cattle manure (45 kg), and lime (0%). Lime 1 (control + lime-1%), lime 2 (control + lime-2%), and lime 3 (control + lime-3%).

F5 fraction) due to decomposition of organic matter during the composting process (Smith, 2009). Furthermore, at higher pH, Zn may precipitate with hydroxides, carbonates, phosphates, sulfides, and numerous further anions as well as form complexes with stabilized organic matter (Kumpiene et al., 2008). Wong and Selvam (2006) also observed that the residual fraction of Zn was found to be dominant in the lime-amended composting of sewage sludge. The BF of Zn was decreased from 0.67 to 0.43 and from 0.61 to 0.36 in the pile composting and the horizontal rotary drum rector composting process of water hyacinth with lime addition (Singh et al., 2015; Singh and Kalamdhad, 2016).

Singh et al. (2015) and Singh and Kalamdhad (2016) reported that the F1, F2, F3, and F4 fractions of Cu were reduced in pile composting and horizontal rotary drum rector composting of water hyacinth with the lime addition. The F5 fraction of Cu was enhanced in the lime-amended compost. The higher reduction in the bioavailable fractions of Cu was observed in lime-added compost due to the conversion of easily available fractions into the residual fraction (Qiao and Ho, 1997). The F1 and F2 fractions of Cu were easily released during the composting process and ions were bound to one or more organic functional groups: in particular, carboxylic, carbonyl, and phenolic groups (Nomeda et al., 2008; Zhu et al., 2011).

The F4 and F5 fractions of Cu were dominant in the final composting of water hyacinth lime treatment. Similar results were also reported by other researchers in the composting of sewage sludge (Wong and Selvam, 2006; Zhu et al., 2011). The BF of Cu was decreased from 0.53 to 0.20 and from 0.64 to 0.20 in lime treatment during the pile composting and horizontal rotary drum rector composting process of water hyacinth (Singh et al., 2015; Singh and Kalamdhad, 2016). The higher reduction in the BF of Cu was observed in lime treatment which might be due to conversion of most bioavailable fractions into the more stable fraction (the F5 fraction). Wong and Selvam (2006) observed that the sequence of Cu fractions was as follows: F4 > F5 > F3 > F1 > F2 in the lime-modified compost of sewage sludge and F5 > F4 > F1 > F2 > F3 in the final compost of water hyacinth with lime (Singh et al., 2015; Singh and Kalamdhad, 2016).

Singh and Kalamdhad (2016) reported that the F1 fraction of Mn was increased in lime treatment. The F2 and F4 fractions of Mn were reduced in lime treatment. The F3 fraction of Mn was reduced approximately 2.9% of total fraction in lime treatment. The F5 fraction of Mn was increased in the lime treatment during the composting process. Wong and Selvam (2006) reported that initially Mn was dominated by the F5 fraction but when the composting progressed, the F3 fraction increased and became the largest fraction. However, in the water composting with lime study, the F3 fraction was found to be predominant during the composting process. Singh et al. (2015) reported that the F1 and F4 fractions of Mn were reduced in lime treatment. The F1 and F2 fractions were reduced in lime treatment (64.7%) during the composting process, whereas the F3 fraction of Mn was increased. The F5 fraction was increased in the lime treatment during the composting process. Wong and Selvam (2006) reported that the order of different fractions of Mn was as follows: F3 > F5 > F1 > F2 > F4 in control and lime-amended composting of sewage sludge and F5 > F3 > F2 > F1 > F4 in the final compost of water hyacinth with lime. These sequences reveal that the F2 fraction was dominant followed by the F5 fraction in control, whereas in the lime-treated compost, the F3 fraction was found to be dominant followed by the F5 fraction. It can be concluded that the F2 fraction converted into the F3 and F5 fractions due to lime addition. The BF of Mn was decreased from 0.91 to 0.71 in lime treatments during the pile composting, whereas it was decreased from 0.86 to 0.55 during the horizontal rotary drum rector composting process.

Singh et al. (2015) and Singh and Kalamdhad (2016) reported that all the mobile fractions (the F1, F2, F3, and F4 fractions) of Fe were decreased in the lime-treated compost of water hyacinth, whereas the F5 fraction of Fe was increased in lime-amended compost. A significant decrease in the F1 fraction of Fe was found in lime treatment, suggesting that the lime-amended compost could check the mobility of Fe. The F4 fraction of Fe was found to be dominant at the initial stage of composting, whereas its F5 fraction was dominant at the end of composting process. It might be, due to the addition of lime, that the F4 fraction was converted into the F5 fraction. The order of different fractions of Fe was as follows: F5 (67.6%) > F4 (15.8%) > F3 (11.5%) > F1 (1.4%) > F2 (0.15%). The BF of Fe was decreased from 0.82 to 0.29 and from 0.75 to 0.30 in the pile composting and horizontal rotary drum reactor composting process, respectively (Singh et al., 2015; Singh and Kalamdhad, 2016).

Table 7.4 shows the speciation of Ni, Pb, Cd, and Cr in lime treatment during agitated pile composting and rotary drum of composting of water hyacinth. The F1 and F2 fractions of Ni were reduced in the lime treatment, whereas the F5 fraction of Ni was enhanced in the lime treatment during the composting process (Singh et al., 2015; Singh and Kalamdhad, 2016). The F1 fractions contributed <1.5% of total fraction in lime-treated compost, whereas the F2 fractions contributed <1% of the

TABLE 7.4
Speciation of Ni, Pb, Cd, and Cr in Lime Treatment during Agitated Pile Composting and Rotary Drum Composting of Water Hyacinth

		Speciation of Heavy Metals									
		Ni (mg/kg)					Pb (mg/kg)				
Composting Methods	Days	F1	F2	F3	F4	F5	F1	F2	F3	F4	F5
Agitated pile	0	2.58	3.25	ND	ND	210.8	24.9	10.8	ND	ND	803
	30	1.15	1.20	ND	ND	337.8	16.0	7.5	ND	ND	1153
Rotary drum	0	1.50	3.97	ND	ND	191.4	16.15	14.20	ND	ND	683.4
	20	1.00	1.00	ND	ND	252.5	5.43	8.07	ND	ND	897.5
		Cd (mg/kg)					Cr (mg/kg)				
Agitated pile	0	0.94	1.25	ND	ND	40.85	3.75	2.29	1.75	11.0	97.9
	30	0.48	0.79	ND	ND	64.8	1.80	1.18	1.15	7.0	157.5
Rotary drum	0	0.87	0.86	ND	ND	37.5	1.5	2.57	3.20	4.68	78.5
	20	0.33	0.53	ND	ND	56.8	0.79	1.50	1.55	1.75	139.4

Note: Composition of treated compost: water hyacinth (90 kg), sawdust (15 kg), cattle manure (45 kg), and lime (0%). Lime 1 (control + lime-1%); lime 2 (control + lime-2%), and lime 3 (control + lime-3%).

ND, not detected.

total fraction in lime-treated compost. Wong and Selvam (2006) reported that both the F1 and F2 fractions contribute approximately 2% of the total Ni contents. The F5 fraction of Ni was approximately 98% in the lime-treated compost, whereas the F3 and F4 fractions of Ni were not detected throughout the composting process. A significant decrease in BF was observed in lime treatment, suggesting that the addition of lime could prevent the bioavailability of Ni during the process. Wong and Selvam (2006) reported that the F5 fraction of Ni was found to be dominant followed by F3, F4, F2, and F1 in lime-treated compost of sewage sludge. However, Zhu et al. (2011) reported that Ni was primarily distributed in the F3 and F4 fractions in sewage sludge amended in soil. The order of different fractions of Fe in the lime-treated composting of water hyacinth was as follows: F5 > F2 > F1. The BF of Ni was decreased from 0.027 to 0.007 and from 0.028 to 0.008 in the APC and RDC of water hyacinth with lime treatment (Singh et al., 2015; Singh and Kalamdhad, 2016). Su and Wong (2003) reported that the F5 fraction of Ni contributed 52% of the total Ni content followed by the F3 fraction in sewage sludge, whereas the F5 fraction of Ni contributed approximately 96%–98% of the total fractions in the water hyacinth composting.

Singh et al. (2015); Singh and Kalamdhad (2016) reported that the exchangeable and carbonate fractions of Pb were reduced, whereas its residual fraction was increased in the lime-treated compost of water hyacinth. Wong and Selvam (2006) reported that there is no marked difference in the F1 and F2 fractions during the composting of sewage sludge. According to Singh et al. (2015) and Singh and Kalamdhad (2016), the F5 fraction of Pb was found to be dominant in the final compost of lime treatment. The development of slightly alkaline medium by lime addition could decrease mobility of Pb due to formation of Pb-organic matter composite; however, it has an inverse effect on Pb composite with organic matter because of its the amphoteric characteristic at alkaline conditions (Kumpiene et al., 2008). Singh et al. (2015) and Singh and Kalamdhad (2016) also reported that the order of different fractions was as follows in lime-treated compost: F5 > F2 > F1. The F3 and F4 fractions of Pb were not detected in the final composting of water hyacinth. The BF of Pb reduced from 0.042 to 0.020 and from 0.042 to 0.015 in the pile composting and horizontal rotary drum reactor composting process of water hyacinth composting with lime. Qiao and Ho (1997) also stated that the mobility of Pb was decreased in the composting of sewage sludge.

Singh et al. (2015); Singh and Kalamdhad (2016) reported that the mobile forms (the F1 and F2 fractions) of Cd were decreased significantly in the lime-treated compost of water hyacinth prepared in the pile composting and horizontal rotary drum reactor. The F1 and F2 portions were decreased and the F5 fraction was increased due to formation of strong chemical bonds between Cd and degraded organic matter (Haroun et al., 2007). A similar observation was also reported by Hanc et al. (2009) during the composting of sewage sludge. Liu et al. (2007) reported that the order of Cd fractions was as follows: F3 > F2 > F4 > F1 > F5in the composting of sewage sludge, whereas this order was: F5 > F2 > F1 in lime-treated composting of water hyacinth. The F3 and F4 fractions of Cd were not detected (Singh et al., 2015); Singh and Kalamdhad, 2016). The BF of Cd was decreased from 0.051 to 0.019 and from 0.044 to 0.015 in the pile composting and horizontal rotary drum reactor composting process of water hyacinth composting with lime. This is likely due to neutral or alkaline pH in lime treatment during the composting process (Singh et al., 2015; Singh and Kalamdhad, 2016).

The F1, F2, F3, and F4 fractions of Cr were reduced in the lime treatment. The F5 fraction of Cr was enhanced in lime-amended compost and this fraction was dominant throughout the composting process. A significant decrease in the BF was detected in lime treatment due to the binding of the F1 and F2 forms with numerous organic functional groups present in the humus-like materials, and the F3 and F4 fractions were transformed into the F5 fraction at the maturity of composting (Singh et al., 2015; Singh and Kalamdhad, 2016).

Smith (2009) also reported that Cr was mainly present in the F5 fraction. The order of different fractions of Cr was as follows: F5 > F4 > F3 > F2 > F1. However, Liu et al. (2007) described that the order of various fractions of Cr in the composting of sewage sludge was as follows: F4 > F5 > F1 = F3 > F2. The BF of Cr was decreased from 0.16 to 0.07 and from 0.13 to 0.04 in the pile composting and horizontal rotary drum reactor composting process of water hyacinth with lime, respectively (Singh et al., 2015; Singh and Kalamdhad, 2016).

7.3 EFFECTS OF NATURAL ZEOLITE ON BIOAVAILABILITY AND SPECIATION OF HEAVY METALS

According to the study of Zorpas et al. (2002), the metal uptake on clinoptilolite decreases with the decreasing particle size of clinoptilolite,due to surface dust on it, which clogs pores and causes structural damage in smaller particles during to the grinding process. Sprynskyy et al. (2007) reported that anaddition of the natural clinoptilolite to the sewage sludge, directed to the fractions of metals decreased according to sequence of the sequential extraction procedure. The fractions of metals (Cd, Cr, Cu, and Ni) decreased by 87% for Cd, 64% for Cr, 35% for Cu, and 24% for Ni due to addition of the clinoptilolite to the composting mixture. The total concentrations of metals were reduced by 11%, 15%, 25%, 41%, and 51% for Cu, Ni, Cr, Cd, and Pb, respectively, due to addition of clinoptilolite to the sludge. The clinoptilolite rock may be considered a suitable material for the immobilization of heavy metals.

Stylianou et al. (2008) revealed that the hydration of the cation is also important during the process; for example, in clinoptilolite-water ion exchange systems the series is as follows: $Pb^{2+} > Cr^{3+} > Fe^{3+} > Cu^{2+}$ and $Pb^{2+} > Zn^{2+} > Cu^{2+}$. It should be noted that organic matter in soluble and insoluble forms plays contrasting roles in controlling total soluble metals. It promotes the dissolution of Cu, Zn, and other metals by building organic complexes since the addition of organic matter increased the solubility of metals by the formation of organometallic complexes (Stylianou et al., 2008).

7.3.1 EFFECTS OF NATURAL ZEOLITE ON WATER SOLUBILITY OF HEAVY METALS

Table 7.5 displays the variations in water solubility of heavy metals in water hyacinth composting with zeolite addition. The water solubility of metals such as Zn, Cu, Mn, and Cr was reduced approximately 80%, 76.7%, 83.1%, and 100%, respectively, in

TABLE 7.5
Influence of Natural Zeolite on Water Extractability of Heavy Metals in the Pile Composting and Horizontal Rotary Drum Reactor Composting of Water Hyacinth

		Water-Extractable Metals Concentration (mg/kg)							
Composting Methods	**Days**	**Zn**	**Cu**	**Mn**	**Fe**	**Ni**	**Pb**	**Cd**	**Cr**
Agitated pile	0	1.49	1.77	3.80	9.11	ND	ND	ND	0.62
	30	0.46	0.49	1.11	3.65	ND	ND	ND	0.0
Rotary drum	0	1.57	1.83	3.34	31.09	ND	ND	ND	1.73
	20	0.73	0.435	0.97	10.8	ND	ND	ND	0.33

ND, not detected.

the agitated pile composting process (Singh and Kalamdhad, 2014b). However, the water extractability of Zn, Cu, Mn, Fe, and Cr were decreased (percentage of total metals) approximately 71.3%, 79.1%, 78.3%, 76.8%, and 89.4%, respectively, during the horizontal rotary drum reactor composting of water hyacinth (Singh et al., 2013). The waste solubilities of metals (Zn, Cu, Mn, Fe, and Cr) were found in 5% and 10% zeolitetreatments. The water-extractable fractions of Cd, Ni, and Pb were not found during the agitated pile and rotary drum composting. Stylianou et al. (2008) stated that water-soluble fractions of metals were released during degradation of organic matter in the composting of sewage sludge, therefore more heavy metals are available in F1 fraction, and finally this fraction was bind with zeolite by ion exchange route.

7.3.2 Effects of Natural Zeolite on DTPA Extractability

The DTPA-extractable fractions of Pb, Cu, and Zn expressively reduced in the composting due to the addition of zeolite. It has high ion exchange capacity with metals (Chiang et al., 2007). The DTPA extraction efficiency of metals was decreased in the agitated pile and rotary drum composting of water hyacinth with natural zeolite (Table 7.6).The DTPA extraction of heavy metals was decreased in horizontal rotary drum reactor composting in comparison to the agitated pile composting (Singh et al., 2013; Singh and Kalamdhad, 2014b). The DTPA-extractable concentrations of Pb and Cd were not detected during agitated pile and rotary drum composting of water hyacinth. Chiang et al. (2007) also reported the decrease in concentration of metals extracted with DTPA in the composting of sewage sludge with the addition of natural zeolite. This decrease in metals could be ascribed as metal ions exchanged with Na, K, and Ca ions in the composting process (Zorpas et al., 2000; Erdem et al., 2004).

Garcia et al. (1995) stated that DTPA extraction of metals was decreased due to the conversion of organic matter into humus at the maturity of the composting, leading to the formation of metal–humus complex (Garcia et al., 1995). Xiong et al. (2010) reported that ligneous bulking material particularly wood sawdust encourage composting process resulting more stabilized material is formed to which metal bind

TABLE 7.6
Effects of Natural Zeolite on DTPA Extractability of Heavy Metals in the Agitated Pile Composting and Horizontal Rotary Drum Reactor Composting Water Hyacinth

		DTPA Extraction Concentration of Metals (mg/kg)							
Composting Methods	**Days**	**Zn**	**Cu**	**Mn**	**Fe**	**Ni**	**Pb**	**Cd**	**Cr**
Agitated pile	0	36.16	11.39	151.60	81.05	0.86	ND	ND	0.62
	30	18.22	3.69	83.29	41.63	0.24	ND	ND	0.33
Rotary drum	0	24.80	7.93	174.75	106.95	0.74	ND	ND	2.73
	20	13.435	1.84	130.30	88.25	0.058	ND	ND	0.83

ND, not detected.

up strongly consequently bioavailability of heavy metals decreased, this compost is safe and can be applied to land.

7.3.3 Effect of Natural Zeolite on Speciation of Heavy Metals

Speciation of heavy metals (Zn, Cu, Mn, and Fe) in the pile composting and horizontal rotary drum reactor composting of water hyacinth with natural zeolite is given in Table 7.7. All the mobile fractions (the F1, F2, F3, and F4 fractions) of Zn were decreased approximately 4.9% of total fractions in zeolite treatment during the water hyacinth composting. The F1 and F2 fractions of Zn are the most movable and bioavailable fractions which were reduced in zeolite treatment during water hyacinth

TABLE 7.7
Speciation of Zn, Cu, Mn, and Fe in Pile and Horizontal Rotary Drum Reactor Composting of Water Hyacinth with Natural Zeolite

		Speciation of Heavy Metals									
		Zn (mg/kg)					Cu (mg/kg)				
Composting Methods	**Days**	**F1**	**F2**	**F3**	**F4**	**F5**	**F1**	**F2**	**F3**	**F4**	**F5**
Agitated pile	0	8.4	7.7	29.0	10.1	203	4.8	1.8	1.7	7.3	131.2
	30	2.5	2.6	20.9	12.9	257	1.3	0.4	0.4	3.87	161.8
Rotary drum	0	8.2	7.6	26.1	15.7	147	2.3	2.1	2.5	10.9	65.1
	20	5.2	3.2	18.0	11.2	227	1.1	0.9	1.3	9.24	86.2
		Mn (mg/kg)					Fe (mg/kg)				
Agitated pile	0	133	76.8	76.8	46.0	92.2	10.7	7.80	1210	6156	12450
	30	67.5	35.3	101	54.3	475	4.5	3.6	893	3089	23528
Rotary drum	0	122	83.2	129	83.1	71.3	19.2	31.5	1953	2686	14563
	20	64.6	71.4	83.5	69.2	280	13.3	17.5	1440	2361	26048

composting with zeolite. The F5 fraction of Zn was increased in the final compost of water hyacinth. The order of different fractions of Zn in the zeolite treatment was as follows: F5 > F3 > F4 > F2 > F1. The BF of Zn was decreased from 0.21 to 0.13 in zeolite treatment during the agitated pile composting of water hyacinth. The BF of Zn was decreased from 0.28 to 0.14 in zeolite treatment in the rotary drum composting. The higher reduction of the BF in zeolite treatments might be due to the transformation of organic matter into humic substances which formed complexes with Zn through its various organic functional groups (Cai et al., 2007). Zorpas et al. (2008) stated that the F3 and F4 fractions of Zn were found to be dominant in the natural zeolite-amended sewage sludge composting, whereas its F5 fraction was increased significantly in the water hyacinth compost treated with zeolite. The decrease of the BF in zeolite treatment can be attributed to the fact that Zn formed complex compounds with many organic functional groups present in humus (Cai et al., 2007). Kumpiene et al. (2008) recommended that complexation of Zn with organic legends as well as cation exchange are the main mechanisms of controlling bioavailability of Zn. The F1, F2, F3, and F4 fractions of Cu were reduced in the control and all zeolite treatments in agitated pile and rotary drum composting.

According to Zorpas et al. (2000), the F4 and F5 fractions of Cu were increased significantly in the composting of sewage sludge with natural zeolite, whereas the F1 and F2 fractions of Cu contributed only 2% of total Cu concentration. Singh and Kalamdhad (2014c, d) reported that the F5 fraction of Cu was found to be dominant, most mobile fractions (the F1 and F2 fractions) were found to be <2% in both pile and horizontal rotary drum reactor composting of water hyacinth with zeolite treatment. This study also reported that all mobile fractions (F1, F2, F3, and F4) were decreased, whereas the F5 fraction was increased in the composting of water hyacinth treated with zeolite. Mobile fractions of Cu are generally converted into the residual fraction due to the binding of mobile fractions with carboxyl groups of humic substances (Liu et al., 2008). The residual fraction of Cu was mainly found in in the zeolite-treated compost of water hyacinth. However, Zorpas et al. (2000) reported that the oxidizable and residual fractions of Cu were found mainly in the composting of sewage sludge with amendment of natural zeolite. The BF of Cu was decreased from 0.11 to 0.04 in zeolite treatment during the agitated pile composting of water hyacinth, whereas the BF of Cu was decreased from 0.22 to 0.13 in the zeolite-treated composting of rotary drum. The decrease in BF of Cu in zeolite treatment can be explained as higher degradation of organic matter during composting process followed by conversion of it into humic substances causing reduction in mobile fractions or converting into residual fraction of Cu at the maturity of water hyacinth compost. Humic substance formed during the composting process has potential to absorb Cu strongly (Farrell and Jones, 2009). The humification of organic matter during composting may be accountable for the decrease in the exchangeable and reducible fractions and the increase in the organically bound fraction (Ingelmo et al., 2012).

Singh and Kalamdhad (2014c,d) reported that all the mobile fractions (the F1, F2, F3, and F4 fractions) of Mn were decreased in the agitated pile and rotary drum composting of water hyacinth with zeolite. The F1 and F2 fractions contributed approximately 23.9% of the total fraction in the zeolite treatment in the horizontal rotary drum reactor of composting of water hyacinth. The F5 fraction of Mn was increased due to conversion of all the mobile fractions (F1 to F4) into the highly stable fraction (the F5 fraction) in the composting of water hyacinth. The F5 fraction of Mn was increased from 36.7% to 66% of the total fractions in the final composting with and without zeolite treatments, whereas Zorpas et al. (2000) described that the F3 fraction of Mn contributed approximately 60% of the total fraction in the final composting of sewage sludge treated with natural zeolite. The BF of Mn reduced from 0.78 to 0.35 in the pile composting of water hyacinth with zeolite treatment, whereas the BF of Mn was decreased from 0.81 to 0.50 in the horizontal rotary drum reactor composting with zeolite treatment.

Singh and Kalamdhad (2014c,d) reported that all the mobile fractions (F1, F2, F3, and F4) of Fe were decreased in both the pile and horizontal rotary drum reactor composting of water hyacinth with zeolite addition. Higher reduction of the F1 and F2 fractions were observed: approximately 70.9% and 74.5% of the total fraction in zeolite treatment. The F1 and F2 fractions contributed approximately 0.10% of the total fraction in zeolite treatment in horizontal rotary drum reactor composting of water hyacinth composting. The F5 fraction of Fe was increased in control and all zeolite treatments during the agitated pile and rotary drum composting of water hyacinth composting. The F5 fraction contributed approximately 87.2% of the total fraction in zeolite treatment during rotary drum composting process. Zorpas et al. (2000) informed that the F3 and F5 fractions of Fe contributed approximately 95% of the total fractions in the water hyacinth composting, whereas its F5 fraction was found mainly in the water hyacinth composting. Zorpas et al. (2008) provided a sequence of difference fractions of Fe in the compost of sewage sludge treated with zeolite. This study also reported that the F5 fraction contributed the highest amount of approximately 64.2% followed by the F3, F4, F2, and F1 fractions. The F5 fraction of Fe contributed approximately 86% followed by the F4, F3, F1, and F2 fractions in the water hyacinth composting. The F5 fraction of Fe was found to be dominant in zeolite-treated compost (Singh and Kalamdhad, 2014c, d). Zorpas et al. (2000) reported that the F3 and F5 fractions of Fe contributed approximately 95% of the total fractions. The BF of Fe was decreased from 0.37 to 0.15 and 0.24 to 0.13 in the pile and horizontal rotary drum reactor composting of water hyacinth with zeolite, suggesting that amendment of zeolite could inhibited the bioavailability of Fe (Singh and Kalamdhad, 2014c,d).

The speciation of heavy metals (Cd Cr, Ni, and Pb) in pile composting and horizontal rotary drum reactor composting of water hyacinth with natural zeolite is specified in Table 7.8. The most mobile fractions of Ni (the F1 and F2 fractions) were decreased in the zeolite treatment in both the agitated pile and rotary drum composting. These fractions were reduced in the range of 67%–71.5% of the total fraction of Ni in zeolite treatment compost prepared using the pile and horizontal rotary drum reactor composting methods (Singh and Kalamdhad, 2014c,d). Zeolites have the capacity to uptake the F1 and F2 fractions of Ni (Zorpas et al., 2008). Singh and Kalamdhad (2014c,d) reported that the F1 and F2 fractions contribute approximately 3% and 1.8% of the total fraction of Ni in pile composting and horizontal rotary drum reactor composting, respectively. The F5 fraction of Ni was enhanced in the zeolite treatment during the agitated pile and rotary drum composting. The F5

TABLE 7.8
Speciation of Ni, Pb, Cd, and Cr in Pile Composting and Horizontal Rotary Drum Reactor Composting of Water Hyacinth with Natural Zeolite

		Speciation of Heavy Metals									
		Ni (mg/kg)					Pb (mg/kg)				
Composting Methods	**Days**	**F1**	**F2**	**F3**	**F4**	**F5**	**F1**	**F2**	**F3**	**F4**	**F5**
Agitated pile	0	8.2	4.9	ND	ND	173	22.9	10.3	ND	ND	792.5
	30	4.0	2.0	ND	ND	220	10.0	4.0	ND	ND	1044
Rotary drum	0	6.7	3.5	ND	ND	210	12.0	11.9	ND	ND	940
	20	2.5	2.2	ND	ND	254	5.0	4.9	ND	ND	1193
		Cd (mg/kg)					Cr (mg/kg)				
Agitated pile	0	0.9	0.7	ND	ND	48.2	3.7	1.7	2.1	0.6	65.1
	30	0.4	0.5	ND	ND	67.5	2.4	0.9	1.05	2.7	112.8
Rotary drum	0	1.3	0.9	ND	ND	71.3	2.4	0.0	2.43	11.3	59.5
	20	0.7	0.2	ND	ND	85.5	1.3	0.0	1.0	3.8	86.0

ND, not detected.

fraction contributed approximately 93%–99.15% of the total fraction in the zeolite treatment during the agitated pile and rotary drum composting of water hyacinth (Singh and Kalamdhad, (2014c, d). A significant increase in the F5 fraction of Ni may be due to the alkaline stabilization process (Gupta and Sinha, 2007). Zorpas et al. (2008) reported that Ni was found to be associated with the F3 fraction (36%) and the F5 fraction (23%) during the composting of sewage sludge with natural zeolite.

Singh and Kalamdhad (2014c,d) reported that the F3 and F4 fractions were not detected throughout the composting process of water hyacinth. The BF of Ni decreased from 0.07 to 0.03 and 0.046 to 0.018 in the agitated pile and rotary drum composting, respectively, due to the bioavailability of Ni by addition of zeolite during the composting process in the zeolite-treated composting. The reduction in BF of Ni may be due to absorption and exchange capacity of zeolite with easily available fractions, mobile fractions bind up with zeolite and also these fractions were transformed into the residual fraction more stable fractions (Zorpas et al., 2008).

Singh and Kalamdhad (2014c,d) reported that the F5 fraction of Pb contributed approximately 98% of total Pb at maturity of the composting process of water hyacinth treated with zeolite. The F3 and F4 fractions were not observed in the final composting. The BF of Pb reduced from 0.040 to 0.013 in zeolite treatment during the pile composting of water hyacinth. However, in the horizontal rotary drum reactor composting of water hyacinth, the BF of Pb was decreased from 0.025 to 0.008 in zeolite treatment. This reduction confirmed that zeolite was efficient for reducing bioavailability of Pb in the mature compost of water hyacinth. The decrease in BF of Pb in the compost treated with zeolite can be attributed the advanced degradation of OM followed by converting OM into humic substances, bioavailable fraction bindup with humic substances, and also accessible fractions were taken up from the zeolite (Zorpas et al., 2008). According to Zorpas et al. (2000), the F3 and F5 fractions of Pb contributed approximately 83% of the total fractions in the composting of sewage sludge with zeolite amendments.

Singh and Kalamdhad (2014c,d) reported that the most mobile fractions (the F1 and F2 fractions) of Cd were decreased and contributed <1% of total Cd, whereas its F5 fraction was increased in the zeolite treatment during the pile composting and horizontal rotary drum reactor composting process of water hyacinth. The BF of Cd was decreased from 0.033 to 0.014 in zeolite treatment during the pile composting of water hyacinth. However, the BF of Cd was decreased from 0.030 to 0.011 in zeolite treatment in the horizontal rotary drum composting of water hyacinth. The reduction of the BF in zeolite treatments might be due to higher degradation of organic matter, resulting in a conversion into humus like substances which formed a metal-humus complex during composting (Tiquia et al., 1997). A significant decrease in F1 and F2 fractions in compost treated with zeolite may be due to ion-exchange reactions occurring in the microporous minerals of the zeolite (Erdem et al., 2004). Zorpas et al. (2000) reported that the F3 fraction of Cd contributed approximately 60% of total Cd in the composting of sewage sludge with natural zeolite. Reduction in the F1 and F2 fractions of Cd attributed to the fact that oxygen containing functional groups with higher affinities to Cd, such as phenolic and benzene-carboxylic groups, were formed during the composting process (Xiong et al., 2010).

Singh and Kalamdhad (2014c, d) reported the all the mobile fractions (the F1, F2, F3, and F4 fractions) of Cr were decreased, whereas its F5 fraction increased in the zeolite treatment in the pile composting and horizontal rotary drum reactor composting of water hyacinth. The most mobile fractions (the F1 and F2 fractions) of Cr were measured approximately 13.0% and 7.2% of the total Cr in the final compost of the agitated pile and rotary drum, respectively, whereas these fractions were found approximately 4.0% and 1.42% of the total fraction in the final compost of the agitated pile with zeolite and the rotary drum with zeolite, respectively. The BF of Cr was decreased from 0.11 to 0.06 and 0.21 to 0.07 in zeolite treatment during the pile and horizontal rotary drum reactor composting, respectively. The reduction in the BF of Cr in zeolite-treated compost treatments was due to the degradation of organic matter and the formation of humus-like substances that bind up eagerly available fractions (F1 and F2) of Cr. The decrease in readily available fractions of Cr is attributed to the fact that zeolites have potential to absorb metal fractions followed by cation exchange of Cr with Na, K, and Ca ions (Zorpas et al., 2008). Zorpas et al. (2000) described that the F4 and F5 fractions of Cr were found to be higher in the compost of zeolite-amended sewage sludge , whereas the F1 and F2 fractions of Cr were found to be <2% in the compost of zeolite-amended sewage sludge. Singh and Kalamdhad (2014c,d) reported that the F5 fraction of Cr was found mainly in the water hyacinth composting.

7.4 OTHER AMENDMENTS

Qiao and Ho (1997) conducted a study on the distribution of heavy metals in the composting of sewage sludge with red mud. This study concluded that red mud influences the distribution of metals by increasing the pH and accessibility of adsorption sites. Generally, sludge composting with red mud addition decreases the availability of heavy metals. The speciation of Cr, Pb and Zn was much affected as compared to the Ni and Cu by addition of red mud in sludge composting. Wang et al. (2008) studied the chemical speciation of heavy metals during the sewage sludge co-composted with sodium sulfide and lime. According to Wang et al. (2008), the heavy metals can be converted into metal sulfides by the sulfidation reaction when Na_2S is added to sewage sludge. This can be expressed as follows:

$$Me^{2+} + S^{2-} = MeS$$

(Me = metals, S = sulfide).

Lu et al. (2014) studied the speciation of Zn and Cu during the composting of pig manure with rock phosphate. This study also reported that the order of different fractions of Cu in the initial compost mixtures was as follows: reducible > exchangeable > oxidizable > residual. The exchangeable and reducible fractions were decreased, whereas the oxidizable and residual fractions were increased during the composting process. In the final composting of pig manure, the percentage fractionation of Cu was found to be similar in control and rock phosphorus treatments (2.5% and 5.0%), and the sequence was as follows: reducible > oxidizable > exchangeable > residual, whereas in the rock phosphorous treatment of 7.5% the sequence was follows: oxidizable > reducible > exchangeable > residual. Qiao and Ho (1997) reported that more than 80% of Cu is tightly bound to the organic fraction, and it was not changed by the red mud addition. Lu et al. (2014) also reported that the BF of Cu in the rock phosphate treatments was found to be higher than that of the control treatment in the initial stage of composting. In the final compost, the reduction in the BF of Cu was found to be highest in the rock phosphate treatments as compared with the control, this order was: rock phosphate treatment of 7.5% > rock phosphate treatment of 5.0% > rock phosphate treatment of 2.5% > the control treatment. The addition of alkaline material (rock phosphate) could dissolve humic substances due to the increase in the BF of Cu in the initial compost mixture (Lu et al., 2014). Wang et al. (2008) reported that Cu was mainly found in the fraction bound to organic matter and sulfides (77.7%) in the initial stage of the composting process. The order of different fractions of Cu in the sewage sludge was as follows: organic-matter-bound and sulfide fraction (77.7%) > residual (20.9%) > carbonate (0.6%) > exchangeable (0.4%) > reducible fraction (0.4%). However, the organic matter-bound and sulfide fractions of Cu were decreased from 77.7% to 62.9% at the end of the composting process, whereas its residual fraction was increased from 20.9% to 35.5%. This study resolved that the organically bound fraction of Cu was converted generally into a residual fraction at the maturity of compost (Wang et al., 2008). However, the F4 fraction of Cu was increased up to 65.3% in compost treated with the sodium sulfide and lime. The increase in the F4 fraction of Cu was mainly due to conversion of heavy metals into metal sulfides (Wang et al., 2008). The exchangeable, carbonate-bound, and reducible fractions of Cu were enhanced faintly at the end of the composting process without adding sodium sulfide and lime, whereas the addition of these chemicals decreased this conversion (Wang et al., 2008).

Wang et al. (2008) reported that the order of different fractions of Ni in sewage sludge was as follows: residual (50.9%) > the organic-matter-bound and sulfide fractions (29.4%) > carbonate (10.2%) > reducible (7.1%) > exchangeable fraction (2.5%). However, this order was changed at the end of the composting process: residual (69.3%) > the organic-matter-bound and sulfide fractions (24.0%) > carbonate (3.4%) > exchangeable fraction (2.2%) > reducible fraction (1.0%). The exchangeable, carbonate-bound, reducible, organic-matter-bound, and sulfide fractions of Ni were decreased at the maturity of composting.

Qiao and Ho (1997), reported that zinc was found in sufficient amount in pig manure and compost prepared from the pig manure, subsequently Zn^{2+} can be predictable to stay in ionic form in solution on the basis of their redox potential for the redox reaction with other metal ions. Lu et al. (2014) reported that the exchangeable fraction of Zn was found to be dominant in the composting of pig manure. The availability of a high concentration of exchangeable Zn inferred a great threat to the environment because of their high mobility. This study reported that the exchangeable and reducible fractions of Zn contributed >95% of total Zn in the composts, since most of Zn was present in the bioavailable fractions. The oxidizable fraction of Zn in the final composts was found to be lower than that of Cu, showing less-stabile Zn complexes with OM.

Lu et al. (2014) stated that the exchangeable fraction of Zn decreased during the composting process, whereas its reducible fraction was increased. The exchangeable fractions of Zn were reduced by approximately 11%, 15%, 20%, and 8.0% in the control treatment, and 5.0% and 7.5% in the rock phosphorus treatment, respectively, from the initial stage to the final stage of the composting process. However, the reducible fractions of Zn were increased in all treatments during composting. In the final compost, most of Zn was present in exchangeable and reducible fractions which represented that the BF of Zn was not much changed during the composting. This research finally concluded that the compost mixtures amended with rock phosphate (2.5% and 5.0% amendment) had a potential to decrease the exchangeable fraction of Zn and increase the reducible fraction as compared to the control treatment. The composting of pig manure with rock phosphorus converted the exchangeable fraction of Zn into the reducible fraction (Lu et al., 2014). The decrease in the exchangeable fraction of Zn in rock phosphate treatments may be due to the formation of less-soluble Zn phosphates complex (Cajuste et al., 2006; Lu et al., 2014). The decrease in the exchangeable fraction and increase in the reducible fraction of Zn in the composting of pig manure can be explained by the fact that Zn has comparatively high potential to adsorb on the surfaces of Fe and Mn oxides (the reducible form) and an increase in pH during composting (Zheljazkov and Warman, 2004; Lu et al., 2014). Wang et al. (2008) reported that the order of various fractions of Zn in sewage sludge was as follows: carbonate (27.5%) > organic-matter-bound and sulfide fractions (25.2%) > residual (24.4%) > reducible (12.9%) > exchangeable (10.0%). The carbonate fractions of Zn were predominated in sewage sludge, whereas the residual fractions of Zn were found to be dominant in the final compost of sewage sludge. The order of the distribution of different fractions of Zn in the compost was as follows: residual (38.4%) > organic-matter-bound and sulfide (26.8%) > reducible (17.9%) > carbonate (13.3%) > exchangeable fraction (3.51%). This study concluded that the sodium sulfide and lime addition reduced the mobile and easily available fractions of Zn in the final compost of sewage sludge.

The mechanism of heavy-metal stabilization by the phosphate (a part of phosphatic rock) is as follows (Cao et al., 2009):

1. Phosphate encourages heavy-metal adsorption onto the sewage sludge.
2. Heavy metals combine with phosphate to form precipitates and minerals.
3. Phosphate surface is also capable to adsorb heavy metals.

7.5 CONCLUSION

The quick development of thermophilic stage was observed in lime treatment due to intense microbial growth. The water-soluble metals and the DTPA-extractable metals were reduced significantly due to lime addition in the pile and horizontal rotary drum reactor composting of water hyacinth. The lowest water extractability of Fe was detected as compared to the other metals however, its concentration was the highest in the water hyacinth compost. The addition of lime was very effective for the reduction of the bioavailability of water-soluble and DTPA-extractable metals during the composting of water hyacinth, sawdust, and cattle manure. The addition of the appropriate proportion of lime sludge could reduce the bioavailability of heavy metals in the composting of sewage sludge and water hyacinth. The addition of the natural zeolite in composting of water hyacinth and sewage sludge suggestively decreased water and DTPA extraction of heavy metals. The maximum reduction of bioavailability and leachability of heavy metals were observed in zeolite treatment during the agitated pile and rotary drum composting processes of water hyacinth. The BF of heavy metals also decreased suggestively in the zeolite-treated compost of water hyacinth. The highest BF was found for Mn followed by Fe, Zn, Cr, Cu, Ni, Cd, and Pb in the final compost of water hyacinth during pile composting with zeolite treatment. However, in the horizontal rotary drum reactor composting of water hyacinth, the sequence of the BF of various metals at the maturity of compost was as follows: Mn (0.23) > Zn (0.14) > Fe (0.13) > Cu (0.09) > Cr (0.066) > Cd (0.011) > Ni (0.009) > Pb (0.007). The total concentration of Cd, Cr, Cu, Mn, Ni and was lower than that of Pb, whereas the BF of these metals was higher than that of Pb in the final compost of water hyacinth. The addition of the appropriate quantity of natural zeolite was very effective for the reduction of the bioavailability of heavy metals in composting of water hyacinth. On the basis of sequential extraction procedure, it was found that zeolite was successful to convert metal in the residual fraction in the final compost of organic substances. Zeolite had adsorbed the most easily bound fractions (exchangeable and carbonate fractions). The addition of sodium sulfide and lime to the composting biomass can be highly effective for the reduction of the exchangeable and organically bound fractions of metals in the composting of sewage sludge, therefore decreasing the bioavailability and risk of heavy metals.

REFERENCES

Achiba, W.B., Gabteni, N., Lakhdar, A., Laing, G.D., Verloo, M., Jedidi, N., and Gallali, T. 2009. Effects of 5-year application of municipal solid waste compost on the distribution and mobility of heavy metals in a Tunisian calcareous soil. *Agriculture Ecosystem and Environment* 130: 156–163.

Baker, L.R., White, P.M., and Pierzynski, G.M. 2011. Changes in microbial properties after manure, lime, and bentonite application to a heavy metal-contaminated mine waste. *Applied Soil Ecology* 48: 1–10.

Cai, Q.Y., Mo, C.H., Wu, Q.T., Zeng, Q.Y., and Katsoyiannis, A. 2007. Concentration and speciation of heavy metals in six different sewage sludge-composts. *Journal of Hazardous Material* 147: 1063–1072.

Cajuste, L.J., Cajuste, L., Garcia-O, C., and Cruz-D, J. 2006. Distribution and availability of heavy metals in raw and acidulated phosphate rock-amended soils. *Communications in Soil Science and Plant Nutrition* 37: 2541–2552.

Cao, X.D., Wabbi, A., Ma, L., Li, B., and Yang, Y. 2009. Immobilization of Zn, Cu, and Pb in contaminated soils using phosphate rock and phosphoric acid. *Journal of Hazardous Materials* 164(2–3): 555–564.

Castaldi, P., Santona, L., and Melis, P. 2006. Evolution of heavy metals mobility during municipal solid waste composting. *Fresenius Environmental Bulletin* 15(9): 1133–1140.

Chen, Q., Luo, Z., Hills, C., Xue, G., and Tyrer, M. 2009. Precipitation of heavy metals from wastewater using simulated flue gas: Sequent additions of fly ash, lime and carbon dioxide. *Water Resource* 43: 2605–2614.

Chiang, K.Y., Huang, H.J., and Chang, C.N. 2007. Enhancement of heavy metal stabilization by different amendments during sewage sludge composting process. *Journal Environmental Engineering and Management* 17(4): 249–256.

Erdem, E., Karapinar, N., and Donat, R. 2004. The removal of heavy metal cations by natural zeolites. *Journal of Colloid Interface Science* 280: 309–314.

Fang, M., and Wong, J.W.C. 1999. Effects of lime amendment on availability of heavy metals and maturation in sewage sludge composting. *Environmental Pollution* 106: 83–89.

Farrell, M., and Jones, D.L. 2009. Heavy metal contamination of mixed waste compost: Metal speciation and fate. *Bioresource Technology* 100: 4423–4432.

Fuentes, A., Llorens, M., Saez, J., Aguilar, M.I., Soler, A., Ortuno, J.F., and Meseguer, V.F. 2004. Simple and sequential extractions of heavy metals from different sewage sludges. *Chemosphere* 54: 1039–1047.

Gabhane, J., William, S.P.M.P., Bidyadhar, R., Bhilawe, P., Anand, D., Vaidya, A.N., and Wate, S.R. 2012. Additives aided composting of green waste: Effects on organic matter degradation, compost maturity, and quality of the finished compost. *Bioresource Technology* 111: 382–388.

Garau, G., Castaldi, P., Santona, L., Deiana, P., and Melis, P. 2007. Influence of red mud, zeolite and lime on heavy metal immobilization, culturable heterotrophic microbial populations and enzyme activities in a contaminated soil. *Geoderma* 142: 47–57.

Garcia, C., Moreno, J.L., Hernfindez, T., and Costa, F. 1995. Effect of composting on sewage sludges contaminated with heavy metals. *Bioresource Technology* 53: 13–19.

Gupta, A.K., and Sinha, S. 2007. Phytoextraction capacity of the plants growing on tannery sludge dumping sites. *Bioresource Technology* 98: 1788–1794.

Hanc, A., Tlustos, P., Szakova, J., and Habart, J. 2009. Changes in cadmium mobility during composting and after soil application. *Waste Management* 29: 2282–2288.

Haroun, M., Idris, A., and Omar, S.R.S. 2007. A study of heavy metals and their fate in the composting of tannery sludge. *Waste Management* 27: 1541–1550.

Ingelmo, F., Molina, M.J., Soriano, M.D., Gallardo, A., and Lapena, L. 2012. Influence of organic matter transformations on the bioavailability of heavy metals in a sludge based compost. *Journal of Environmental Management* 95: S104–S109.

Khan, M.J., and Jones, D.L. 2009. Effect of composts, lime and diammonium phosphateon the phytoavailability of heavy metals in a copper mine tailing soil. *Pedosphere* 19(5): 631–641.

Kumpiene, J., Lagerkvist, A., and Maurice, C. 2008. Stabilization of As, Cr, Cu, Pb and Zn in soil using amendments—A review. *Waste Management* 28: 215–225.

Liu, Y., Ma, L., Li, Y., and Zheng, L. 2007. Evolution of heavy metal speciation during the aerobic composting process of sewage sludge. *Chemosphere* 67: 1025–1032.

Lu, D., Wang, L., Yan, B., Ou, Y., Guan, J., Bian, Y., and Zhang, Y. 2014. Speciation of Cu and Zn during composting of pig manure amended with rock phosphate. *Waste Management* 34: 1529–1536.

Liu, S., Wang, X., Lu, L., Diao, S., and Zhang, J., 2008. Competitive complexation of copper and zinc by sequentially extracted humic substances from manure compost. *Agricultural Science in China* 7(10): 1253–1259.

Montes-Hernandez, G., Concha-Lozano, N., Renard, F., and Quirico, E. 2009. Removal of oxyanions from synthetic wastewater via carbonation process of calcium hydroxide: Applied and fundamental aspects. *Journal of Hazardous Materials* 166: 788–795.

Nomeda, S., Valdas, P., Chen, S.Y., and Lin, J.G. 2008. Variations of metal distribution in sewage sludge composting. *Waste Management* 28: 1637–1644.

Qiao, L., and Ho, G. 1997. The effects of clay amendment and composting on metal speciation in digested sludge. *Water Resource* 31(5): 951–964.

Singh, J., and Kalamdhad, A.S. 2013. Effects of lime on bioavailability and leachability of heavy metals during agitated pile composting of water hyacinth. *Bioresource Technology* 138: 148–155.

Singh, J., Prasad, R., and Kalamdhad, A.S. 2013. Effect of natural zeolite on bioavailability and leachability of heavy metals during rotary drum composting of water hyacinth. *Research Journal of Chemistry and Environment* 17(8): 26–34.

Singh, J., and Kalamdhad, A.S. 2014a. Effects of carbide sludge (lime) on bioavailability and leachability of heavy metals during rotary drum composting of water hyacinth. *Chemical Speciation and Bioavailability* 26(2): 76–84.

Singh, J., and Kalamdhad, A.S. 2014b. Uptake of heavy metals by natural zeolite during agitated pile composting of water hyacinth. *International Journal of Environmental Science* 5(1): 1–15.

Singh, J., and Kalamdhad, A.S. 2014c. Effects of natural zeolite on speciation of heavy metals during agitated pile composting of water hyacinth. *International Journal of Recycling of Organic Waste in Agriculture* 3(55): 1–17.

Singh, J., and Kalamdhad, A.S. 2014d. Influences of natural zeolite on speciation of heavy metals during rotary drum composting of green waste. *Chemical Speciation and Bioavailability* 26(2): 65–75.

Singh, J., Kalamdhad, A. S., and Lee, B. K. 2015. Reduction of eco-toxicity risk of heavy metals in the rotary drum composting of water hyacinth: Waste lime application and mechanisms. *Environmental Engineering Research* 20(3): 212–222.

Singh, J., and Kalamdhad, A.S. 2016. Effect of lime on speciation of heavy metals during agitated pile composting of water hyacinth. *Frontiers of Environmental Science and Engineering* 10(1): 93–102.

Smith, S.R. 2009. A critical review of the bioavailability and impacts of heavy metals in municipal solid waste composts compared to sewage sludge. *Environment International* 35: 142–156.

Sprynskyy, M., Kosobucki, P., Kowalkowski, T., and Buszewsk, B. 2007. Influence of clinoptilolite rock on chemical speciation of selected heavy metals in sewage sludge. *Journal of Hazardous Materials* 149: 310–316.

Stylianou, M.A., Inglezakis, V.J., Moustakas, K.G., and Loizidou, M.D. 2008. Improvement of the quality of sewage sludge compost by adding natural clinoptilolite. *Desalination* 224: 240–249.

Su, D.C., and Wong, J.W.C. 2003. Chemical speciation and phytoavailability of Zn, Cu, Ni and Cd in soil amended with fly ash-stabilized sewage sludge. *Environment International* 29: 895–900.

Tiquia, S.M., Tam, N.F., and Hodgkiss, I.J. 1997. Effects of turning frequency on composting of spent pig-manure sawdust litter. *Bioresource Technology* 62: 37–42.

Villasenor, J., Rodriguez, L., and Fernandez, F.J. 2011. Composting domestic sewage sludge with natural zeolites in a rotary drum reactor. *Bioresource Technology* 102(2): 1447–1454.

Wang, X., Chen, L., Xia, S., and Zhao, J. 2008. Changes of Cu, Zn, and Ni chemical speciation in sewage sludge co-composted with sodium sulfide and lime. *Journal of Environmental Science* 20: 156–160.

Wong, J.W.C., and Selvam, A. 2006. Speciation of heavy metals during co-composting of sewage sludge with lime. *Chemosphere* 63: 980–986.

Xiong, X., Yan-Xia, L., Ming, Y., Feng-Song, Z., and Wei, L. 2010. Increase in complexation ability of humic acids with the addition of ligneous bulking agents during sewage sludge composting. *Bioresource Technology* 101: 9650–9653.

Zheljazkov, V.D., and Warman, P.R. 2004. Phytoavailability and fractionation of copper, manganese, and zinc in soil following application of two composts to four crops. *Environmental Pollution* 131: 187–195.

Zhu, R., Wu, M., and Yang, J. 2011. Mobilities and leachabilities of heavy metals in sludge with humus soil. *Journal of Environmental Science* 23(2): 247–254.

Zorpas, A.A., Constantinides, T., Vlyssides, A.G., Haralambous, I., and Loizidou, M. 2000. Heavy metal uptake by natural zeolite and metals partitioning in sewage sludge compost. *Bioresource Technology* 72: 113–119.

Zorpas, A.A., Vassilis, I., Loizidou, M., and Grigoropoulou, H. 2002. Particle size effects on uptake of heavy metals from sewage sludge compost using natural zeolite clinoptilolite. *Journal of Colloid Interface Science* 250: 1–4

Zorpas, A.A., Vassilis, I., and Loizidou, M. 2008. Heavy metals fractionation before, during and after composting of sewage sludge with natural zeolite. *Waste Management* 28: 2054–2060.

8 Reduction of Bioavailability of Heavy Metals by Microorganisms

8.1 COMPOSTING MICROORGANISMS

Composting is fundamentally a sensation of microbial activity prompting and mainly affected by the thermophilic temperature developed during the process due to microbial degradation and change in different microbial communities (Varma and Kalamdhad, 2014). The standardized microbial examination can be possibly used for refereeing the quality and maturity of compost. However, the impact of temperature and its interface with chemicals are mainly responsible for the succession of microbial communities in the composting process (Zheng et al., 2007). The flexibility of compost system is usually ascribed to the highly active and diverse microbial population (Khan and Joergensen, 2009). A very few knowledges are available on, succession and participation of microbial populations during the specific phases of composting process (Adams and Frostick, 2008). *Phanerochaete chrysosporium*—the descriptive species of white-rot fungi—is a well-known, highly efficient fungus that has the potential to degrade different organic substrates (Huang et al., 2017).

Ahmad et al. (2007) the studied of microorganism in the composting of tannery sludge with sawdust, chicken manure, and rice bran; reported that total aerobic mesophiles (5×10^6 CFUg^{-1} fresh compost), bacilli (8×10^9 CFUg^{-1} fresh compost), *Salmonella spp.* and *Shigella spp.* (<10), and yeasts and molds (4.1×10^6 CFUg^{-1} fresh compost) were present in the composting biomass. The microbial population was significantly decreased at the maturity of the composting period to reach <10 CFUg^{-1} fresh compost total aerobic mesophile, 9.5 $\times 10^2$ CFUg^{-1} fresh compost bacilli, and <10 CFUg^{-1} fresh compost yeasts and molds. However, *Salmonella spp.* and *Shigella spp* were not noticed in the final compost of tannery sludge. Varma and Kalamdhad (2014) studied microbial population in the rotary drum composting of vegetable waste and reported that microbial growth rate was affected by the temperature and also the actual degradation of organic matter in the process. Furthermore, appropriate mixing of various waste materials had a significant role for the succession of microbial populations. Spore-forming bacteria were mainly found during the composting process of the thermophilic stage. A considerable amount of fungi, actinomycetes, and streptomycetes were measured in the final stages of composting process. The populations of fungi, actinomycetes, and streptomycetes represent the presence of lingocellulosic material in the composting biomass. A decrease in the microbial population in the final stage of composting can be ascribed to the reduction availability of the nutrients from the composting biomass. Preserving the temperature at 60°C during the first month of composting caused a significant elimination of the total aerobic mesophile, yeasts, moulds, *Salmonella spp.*, and *Shigella spp.* (Ahmad et al., 2007). Shukla et al. (2009) reported that the microbial population (Figure 8.1) has been changed in microbial biomass during composting of biomass of *V. spiralis*, which has been used for phytoremediation. Vishan et al. (2014) studied total coliform and fecal coliform counts in the initial and final compost of water hyacinth. This study also reported that the fractions of total coliform and fecal coliform were decreased from 1.1×10^{10} to 2.4×10^2 MPN/g and from 1.5×10^6 to 1^5 MPN/g, respectively, during the 20 day composting process of water hyacinth. The *Enterococci* population was detected in the range of 1×10^2 to 2×10^3 CFU/g. The increase of high-degrading temperature during the composting process might have eliminated the growth of these pathogens, resulting in the more stable and hygienic compost. Microbial activities during the composting process play a significant role in determining the biodegradability of organic compounds and for improving the quality of the final compost (Tang et al., 2006). Lignocellulosic materials are present in the composting biomass, which is a significant fraction of the organic matter. Consequently, lignocellulolytic microorganisms are involved

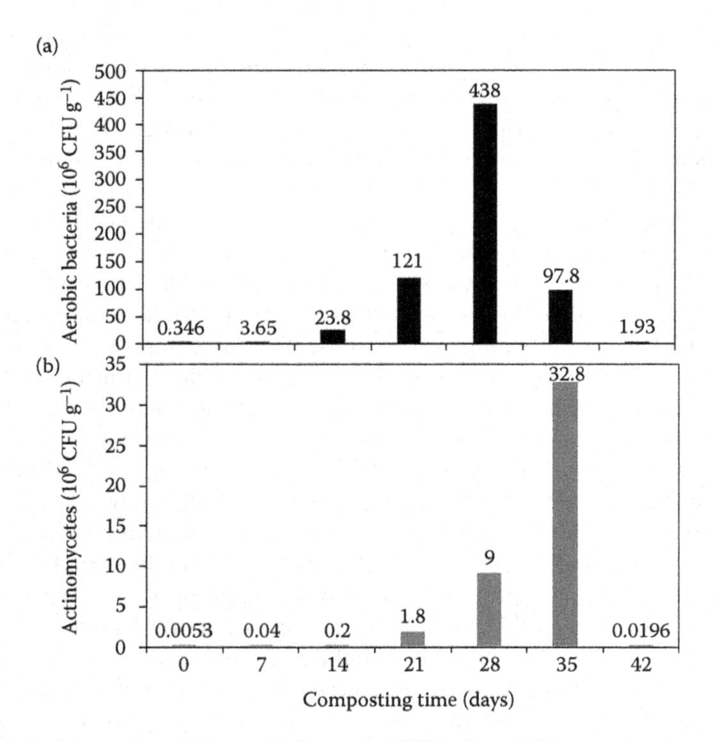

FIGURE 8.1 Interaction of (a) aerobic bacteria (CFU g^{-1}) and (b) actenomycetes (CFU g^{-1}) with composting mixture during the composting process of *Vallisneria spiralis* inside the composter.

in the conversion of lignin and other substances into humus (Tuomela et al., 2000). White-rot fungi, which is a common example of lignocellulolytic microorganism, not only improve the compost quality but also may involve immobilization of heavy metals in the final compost (Zeng et al., 2009; Zhao et al., 2016; Zhang et al., 2018). There are various benefits of using compost in agricultural soils (Wei and Liu, 2005). However, the heavy-metal content is the main factor leading to restricted agricultural use of compost (Wei and Liu, 2005). Conversely, fractionation and chemical speciation of the heavy metals allows the estimation of heavy-metal bioavailability and is related to the different natures of the metals, their bonding strength, either in the free ionic form or complexed by organic matter, or combined with the mineral fraction of the compost (Amir et al., 2005; Gupta and Sinha, 2007).

8.2 REDUCTION OF HEAVY-METAL BIOAVAILABILITY BY MICROORGANISMS

8.2.1 Immobilization of Heavy Metals by Fungi

A very few studies are available on the application of the inoculants of microorganisms mainly fungi to immobilize metals during the composting of different solid wastes which is contaminated with metal (Huang et al., 2006; Zeng et al., 2007, Xu et al., 2014; Zeng et al., 2015; Zhang et al., 2018). Furthermore, the microorganisms are involved in metal adsorption and detoxification of metal-contaminating wastewater. Fungi could be predictable to immobilize metals in the composting process of solid wastes (Zeng et al., 2007). White-rot fungi are useful in absorbing heavy-metal ions from wastewater using their mycelium (Iqbal and Edyvean, 2004; Li et al., 2004; Zeng et al., 2007). This fungus has the potential to accrue heavy metals in its cells by intracellular uptake, and these metal ions are chelated with the carboxyl, hydroxyl, or other active functional groups present in the cell (including the dead cell) surface of microorganisms (Baldrian, 2003; Zeng et al., 2007). Assessment of various forms of Pb (exchangeable, carbonate, reducible, oxidizable and residual) during composting is a dynamic for the development of composting method and the risk valuation of during the compost application to the soil (Zorpas et al., 2003; Huang et al., 2008).

Pb is a metal that is not essential for living organisms. It is a highly toxic metal whose widespread use has caused extensive environmental contamination and health problems in many parts of the world. However, mining and smelting activities, combustion of leaded gasoline, application of sewage sludge to the land, battery disposal, and other Pb-bearing wastes are the primary sources of Pb (Dollar et al., 2001; Huang et al., 2010; Rodrigues et al., 2010).

Microorganisms have the potential to cope with toxic Pb during their growth in the Pb-contaminated substrates and the exposure of microorganisms to metal-contaminated wastes constantly hinders microbial growth and activity (Baldrian et al., 2000; Singhal and Rathore, 2001; Huang et al., 2010). Microbial enzymes may also be affected by heavy-metal toxicity because of their potential inhibition through enzymatic reactions and complex metabolic processes (Tuomela et al.,

2005). Yetis et al. (2000) and Baldrian (2003) reported that *P. chrysosporium* has potential to cope with the Pb(II) concentrations of 5–30 mg/L in aqueous solution, therefore this fungus has been used Pb removal from the wastewater using their mycelium. Huang et al. (2006) and Zeng et al. (2007) reported that organic matter contaminated with Pb (concentrations of 105 and 400 mg/kg dry weight) degraded by *P. chrysosporium* in the process of composting. Huang et al. (2010) reported that under high concentration of Pb *P. chrysosporium* can grow well with good activities of ligninolytic enzymes (LiP and MnP), this study also concluded that under low Pb concentrations activities of ligninolytic enzymes and xylanase were improved, subsequent higher degradation of lignin and hemicellulose degradation was achieved.

Zeng et al. (2007) studied the bioavailability of Pb in the composting of Pb-contaminated solid waste with fungus inocula of white-rot fungus. Results also showed that the phytotoxicity of compost and the bioavailability of Pb decreased with inocula of white-rot fungus in the reactor. Results of this study showed that the composting of Pb-contaminated solid wastes with the inocula of white-rot fungus might improve the microbial activity and reduce the active Pb. Huang et al. (2008) stated that Pb concentrations in the five fractions varied after solid-state fermentation (SSF) of lignocellulosic biomass by white-rot fungi. In all flasks, the concentration of the soluble–exchangeable Pb and the carbonate-bound Pb decreased, whereas those of the other three forms (the reducible, oxidizable and residual forms) were increased after 42 day of SSF. The concentration of the immobilized Pb decreased with an increase of the metal concentrations during the SSF by white-rot fungi. The bioavailability and transfer capability of the five different fractions of Pb decrease in the order of extraction sequence. The ionic form of a metal is usually more toxic to microorganisms as compared with the complexed or absorbed form. Consequently, most of the mobile fractions of Pb were converted into inactive Pb fractions, resulting in the reduction of toxicity and bioavailability of Pb during the SSF by white-rot fungi (Huang et al., 2008) (Table 8.1).

A reduction in bioavailability and toxicity of Pb can be explained as, Pb could be chelate with functional groups (carboxyl, hydroxyl etc.) are present on the surface of white-rot fungi (Baldrian, 2003; Iqbal and Edyvean, 2004) and accelerating the organic matter decomposition by white-rot fungi resulting formation of humus. The chelation of Pb by humus-like material is the mechanism accountable for the reduction of Pb bioavailability (Huang et al., 2006; Zeng et al., 2007).

Liu et al. (2009) studied the behavior of Pb during the composting of Pb-contaminated wastes and concluded that the composting of waste with inoculants of *P. chrysosporium* could effectively transform Pb fractions and decrease the bioavailability of Pb. The results of this study showed that the mobile fraction of Pb was converted into the immobile fraction due to the accumulation of Pb ions by mycelium of fungi as well as formation of complex with humus which is formed at the end the composting process. The concentration of soluble–exchangeable Pb was reduced with the increasing pH and microbial biomass during the composting process. Zhang et al. (2018) also reported that the acid-exchangeable fraction of Pb in the fungal inoculation reactor was higher than control, whereas its mobile fraction was found to be lower than control. The immobilization fraction of Pb in the fungal inoculant reactor was found to be approximately 7% higher than control.

TABLE 8.1
Microorganisms Involved in the Immobilization of Heavy Metals during the Composting of Different Types of Solid Wastes

Micro-Organisms	Name of Fungi/Species	Target Heavy Metals	Composting Materials	References
Fungi	White-rot fungi (*Phanerochaete chrysosporium* strain BKM-F-1767)	Pb	Pb-contaminated solid waste	Zeng et al. (2007)
	White-rot fungi (*Phanerochaete chrysosporium*)	Zn, Cu, Ni, and Pb	Sewage sludge	Zhang et al. (2018)
	White-rot fungi (*Phanerochaete chrysosporium*)	Pb	Used biomass	Zeng et al. (2015)
	White-rot fungi (*Phanerochaete chrysosporium*)	Pb	Lignocellulosic waste	Huang et al. (2008)
	White-rot fungi (*Phanerochaete chrysosporium*)	Pb	Agricultural waste	Huang et al. (2017)
	Trichoderma harzianum and *Phanerochaete chrysosporium*	–	Olive pomace	Haddadin et al. (2009)
Bacteria	*P. azotoformans*	Pb, Cd, Cr, Mn, and Mg	Industrial sludge	Nair et al. (2008)
	Anaerobic, aerobic, nitrifying bacteria, and actinomycetes	Cr	Biomass of *Vallisneria spirates L*	Shukla et al. (2009)
	Bacillus badius AK	Pb	Water hyacinth	Vishan et al. (2017)

Zhang et al. (2018) studied the effects of inoculating white-rot fungi on the mobility of heavy metals and organic matter transformation during the composting process of sewage sludge. This study also reported that the mobile fraction of Zn (the acid-exchangeable and reducible fractions) was increased on day 10 of the composting period in two reactors (control and fungal inoculant), whereas this fraction was reduced significantly at the end of the composting process. The reducible fraction of Zn was decreased in both reactors, whereas the oxidizable fraction of Zn was increased on day 60 of composting period in both the reactors. The oxidizable and residual fractions of Zn in fungal inoculants were found to be higher than those of control. The stabilization rate of Zn in fungal inoculant was 20.31% higher than control (Zhang et al., 2018).

The acid-exchangeable, reducible, and oxidizable fractions of Cu and Ni in fungal inoculation reactor were found to be lower than those of the control. Moreover, the residual fractions of Cu and Ni were found to be 16.69% and 16.86%, respectively, in fungal inoculating reactor as compared with control. The results of this study represented that the mobile fractions of Cu and Ni were converted into the residual fraction due to inoculation of the white-rot fungi in the composting biomass. The white-rot fungi might increase the organic matter mineralization while decreasing the volume of organic matter. The highest degree of stability of composting biomass attained by the white-rot fungi due to the formation of highly stabilized organic matter through decaying lignin (Zhang et al., 2018). Cu has a strong affinity for organic matter and its oxidizable fraction was found to be dominant the in final composting of sewage sludge (Hsu and Lo, 2001; Fuentes et al., 2004). Additionally, humic-like substances formed during the composting process had many functional groups involved in stabilization of Cu in their residual form (Ingelmo et al., 2012).

8.2.2 Immobilization of Heavy Metals by Bacteria

The thermophilic bacteria mainly belonged to the genus of *Bacillus* in the composting of solid waste. *Bacillus stearothermophilus* is the major dominant species at temperature of 65°C, whereas the *Thermus* strain is considered to be useful for the degradation of organic matter during the composting process of thermogenic phase (>70°C) (Fang and Wong, 2000; Singh and Kalamdhad, 2012). Siderophores are chelating agents and are produced from the bacteria. Siderophores are low molecular-weight ligands involved for capturing and supplying iron to support metabolic activity (Nair et al., 2008). Shukla et al. (2009) studied five fractions of chromium during the composting of tannery effluent treated biomass of *Vallisneria spiralis* L. This study also reported that the residual fraction of Cr was found to be higher in the initial stage of the composting process. It has been suggested the residual fraction of Cr was converted into the carbonate fraction in the initial stage of the composting process. The availability of Cr was reduced to due to the formation of metal organic complexes. This study also investigated that Cr was stabilized in the final composting through altering its chemical properties and activity. Composting offers the appropriate environment for the growth of Cr-resistant microbes, which play a key role in the transformation of the mobile fraction into the residual fraction (Shukla et al., 2009).

8.3 MECHANISM OF IMMOBILIZATION OF HEAVY METALS USING MICROORGANISMS

Metal uptake may be possible due to the interaction of physicochemical between the metal and functional groups available on the cell surface of microorganisms during the non-metabolism-dependent biosorption. Physiosorption, ion exchange, and chemisorption are mainly involved in biosorption of metals on the microbial biomass (Ahalya et al., 2003). Although the cell surface of microorganisms consist of mainly polysaccharides, proteins, and lipids (Ahalya et al., 2003; Sardrood et al., 2013), they include many metal-binding groups (e.g., carboxyl, phosphate, amino groups, and sulfate group). However, metal immobilization through non-metabolism-dependent biosorption is comparatively rapid and can be reversible. The mechanism responsible for the removal of heavy-metal toxicity can either be dependent on the cell's metabolism or the area of metal removal which is an independent metabolism, which is called non-metabolism-dependent biosorption and can be classified as extracellular accumulation and precipitation. Figure 8.2 represents the cell surface sorption/precipitation and intracellular accumulation of metals (Davis et al., 2003; Neethu et al., 2015).

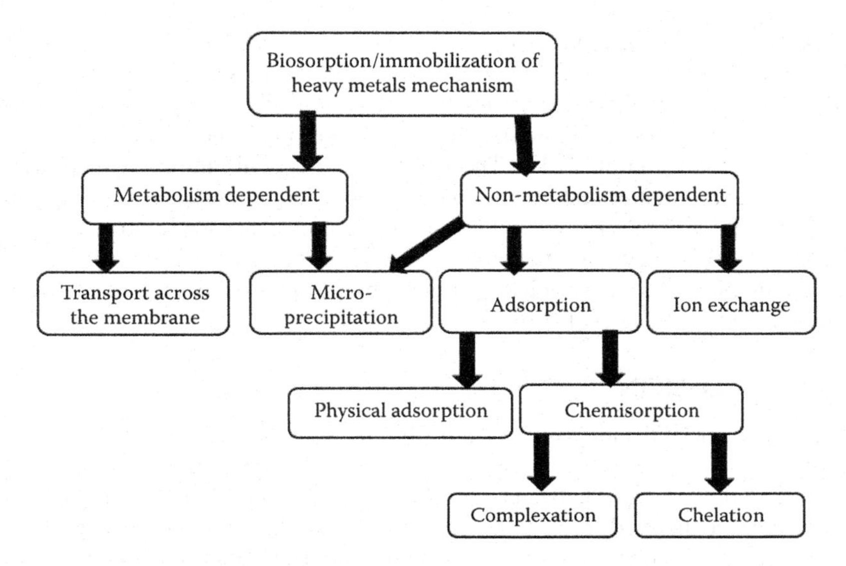

FIGURE 8.2 Mechanisms of biosorption/immobilization of heavy metals by microorganisms.

The following mechanisms are involved for the microbial immobilization/removal of heavy metals:

1. Intracellular sequestration of toxic metals by metal-binding proteins and peptides such as metallothioneins (MTs) and phytochelatins along with complexes produced by bacteria, for example, siderophores that are catecholates that produce hydroxamate-type siderophores. Fungi do not produce siderophores.
2. Modification of biochemical ways to block metal uptake.
3. Transformation of metals into harmless forms through enzymes system.

The mechanisms for detoxifying metals through bacterial action are as follows:

i. Export the toxic ion out of cell by changing a membrane transport system which is initially involved in cellular accumulation.
ii. Reduced permeability.
iii. Extracellular sequestration by specific mineral ion binding (Mustapha and Halimoon, 2015).

Because of the non-biodegradability of heavy metals in the environment, the microbes have defense mechanisms usually based on the stabilization and transformation of toxic metals from one chemical form to another through intracellular and extracellular enzymes of microbes (Yang et al., 2012). There are numerous microorganisms that are capable of secreting extracellular high molecular mass substances. These substances are attached on the cell surface or secrets into the surrounding environment (extracellular polymeric substances, EPS). EPS consists of polysaccharide, protein, lipids, humic substances, and other polymeric complexes. These high molecular-weight secretions from microbes usually contain multiple charged functional groups such as carboxylic, amino, and hydroxylic, which usually possess both adsorptive and adhesive properties resulting in heavy metals immobilized by EPS (Sheng et al., 2013). The white-rot fungi could improve the mineralization of organic matter and sinking the volume of organic matter. Consequently, it could also achieve the advanced stage of composting substances maturity through degradation of lignocellulosic materials resulting in the formation of stable humic-like substances, which had various functional groups to immobile toxic metals as the residual fraction (Zhang et al., 2018).

8.4 CONCLUSION

The use of microbial biomass is one of the most cost-effective methods for the immobilization of heavy metals. These metal ions were significantly affected by the richness and diversity of the bacterial and fungal community, and the mobile fraction of Pb was converted into the inactive fraction of Pb due to the inoculation the metals with *P. chrysosporium*, thus playing an important role in the detoxification of metal ions in microorganisms and the environment. The soluble–exchangeable concentration of Pb was decreased with increasing pH and fungal inoculant. The soluble–exchangeable metal ions are positively correlated with pH value and microbial biomass, indicating that increasing pH and microbial biomass is an important to the immobilization of metal ions during the composting process. Fungi have potential to accumulate metals in their cells by intracellular uptake, and these metal ions are chelated with the carboxyl, hydroxyl, or other active functional groups present in the cell (including the dead cell) surface of microorganisms. Microbial communities such as bacteria, fungi, actinomycetes, etc., play an important role in the detoxification, stabilization, and transformation of metal ions in the composting biomass to ensure the environmental sustainability. Microbial extracellular enzymatic conversion of toxic metal into a less toxic form is the most efficient and cost-effective technology to immobilize metals in the composting of various metal-contaminated wastes.

REFERENCES

Adams, J.D.W., and Frostick, L.E. 2008. Investigating microbial activities in compost using mushroom (*Agaricus bisporus*) cultivation as an experimental system. *Bioresource Technology* 99: 1097–1102.

Ahalya, N., Ramachandra, T., and Kanamadi, R. 2003. Biosorption of heavy metals. *Research Journal of Chemistry and Environment* 7: 71–79.

Amir, S., Hafidi, M., Merlina, G., and Revel, J.C. 2005. Sequential extraction of heavy metals during composting of sewage sludge. *Chemosphere* 59: 801–810.

Baldrian, P., 2003. Interactions of heavy metals with white-rot fungi. *Enzyme and Microbial Technology* 32: 78–91.

Baldrian, P., Wiesche, C., Gabriel, J., Nerud, F., and Zadrazil, F. 2000. Influence of cadmium and mercury on activities of ligninolytic enzymes and degradation of polycyclic aromatic hydrocarbons by *Pleurotus ostreatus* in soil. *Applied and Environmental Microbiology* 66: 2471–2478.

Davis, T.A., Volesky, B., and Mucci, A. 2003. A review of the biochemistry of heavy metal biosorption by brown algae. *Water Research* 37: 4311–4330.

Dollar, N.L., Souch, C.J., Filippelli, G.M., and Mastalerz, M. 2001. Chemical fractionation of metals in wetland sediments: Indiana dunes national lakeshore. *Environmental Science and Technology* 35: 3608–3615.

Fang, M., and Wong, J.W.C. 2000. Changes in thermophilic bacteria population and diversity during composting of coal fly ash and sewage sludge. *Water Air Soil Pollution* 124: 333–343.

Fuentes, A., Llorens, M., Saez, J., Soler, A., Aguilar, M., and Ortuno, F. 2004. Phytotoxicity and heavy metals speciation of stabilized sewage sludges. *Journal of Hazardous Materials* 108: 161–169.

Gupta, A.K., and Sinha, S. 2007. Phytoextraction capacity of the plants growing on tannery sludge dumping sites. *Bioresource Technology* 98: 1788–1794.

Haddadin, M.Y.S., Haddadin, J., Arabiyat, O.I., and Hattar, B. 2009. Biological conversion of olive pomace into compost by using *Trichoderma harzianum* and *Phanerochaete chrysosporium. Bioresource Technology* 100: 4773–4782.

Hsu, J.H., and Lo, S.L. 2001. Effect of composting on characterization and leaching of copper, manganese, and zinc from swine manure. *Environmental Pollution* 114: 119–127.

Huang, D.L. Zeng, G.M., Jiang, X.Y., Feng, C.L., Yu, H.Y., Huang, G.H., and Liu, H.L. 2006. Bioremediation of Pb-contaminated soil by incubating with *Phanerochaete chrysosporium* and straw. *Journal of Hazardous Materials* 134: 268–276.

Huang, D.L., Zeng, G.M., Feng, C.L., Hu, S., Jiang, X.Y., Tang, L., Su, F.F., Zhang, Y., Zeng, W., and Liu, H.L. 2008. Degradation of lead-contaminated lignocellulosic waste by *Phanerochaete Chrysosporium* and the reduction of lead toxicity. *Environmental Science and Technology* 42: 4946–4951.

Huang, D.L., Zeng, G.M., Feng, C.L., Hu, S., Zhao, M.H., Lai, C., Zhang, Y., Jiang, X.Y., and Liu, H.L. 2010. Mycelial growth and solidstate fermentation of lignocellulosic waste by white-rot fungus Phanerochaete chrysosporium under lead stress. *Chemosphere* 81(9): 1091–1097.

Huang, C., Zeng, G.M., Huang, D.L., Lai, C., Xu, P., Zhang, C., Cheng, M., Wan, J., Hu, L., and Zhang, Y. 2017. Effect of *Phanerochaete chrysosporium* inoculation on bacterial community and metal stabilization in lead-contaminated agricultural waste composting. *Bioresource Technology* 243: 294–303.

Ingelmo, F., Molina, M.J., Soriano, M.D., Gallardo, A., and Lapeña, L. 2012. Influence of organic matter transformations on the bioavailability of heavy metals in a sludge based compost. *Journal of Environmental Management* 95: 104–109.

Iqbal, M., and Edyvean, R.G.J. 2004. Biosorption of lead, copper and zinc ions on loofa sponge immobilized biomass of *Phanerochaete chrysosporium. Minerals Engineering* 17: 217–223.

Khan, K.S., and Joergensen, R.G. 2009. Changes in microbial biomass and P fractions in biogenic household waste compost amended with inorganic P fertilizers. *Bioresource Technology* 100: 303–309.

Li, Q.B., Wu, S.T., Liu, G., Liao, X.K., Deng, X., Sun, D.H., Hu, Y.L., and Huang, Y.L. 2004. Simultaneous biosorption of cadmium(II) and lead(II) ions by pretreated biomass of *Phanerochaete chrysosporium. Separation and Purification Technology* 34: 135–142.

Liu, J.X., Xu, X.M., Huang, D.L., and Zeng, G.M. 2009. Transformation behavior of lead fractions during composting of lead-contaminated waste. *Transactions of Nonferrous Metals Society of China* 19: 1377–1382.

Mustapha, M.U., and Halimoon, N. 2015 Microorganisms and biosorption of heavy metals in the environment: A review paper. *Journal of Microbial & Biochemical Technology* 7: 253–256.

Nair, A., Juwarkar, A.A., and Devotta, S. 2008. Study of speciation of metals in an industrial sludge and evaluation of metal chelators for their removal. *Journal of Hazardous Materials* 152: 545–553.

Neethu, C.S., Mujeeb Rahiman, K.M., Saramma, A.V., and Mohamed Hatha, A.A. 2015. Heavy-metal resistance in Gram-negative bacteria isolated from Kongsfjord, Arctic. *Canadian Journal of Microbiology* 61: 429–435.

Rodrigues, S.M., Henriques, B., Coimbra, J., da Silva, E.F., Pereira, M.E., and Duarte, A.C. 2010. Water-soluble fraction of mercury, arsenic and other potentially toxic elements in highly contaminated sediments and soils. *Chemosphere* 78: 1301–1312.

Sardrood, B.P., Goltapeh, E.M., and Varma, A. 2013. *An Introduction to Bioremediation Fungi as Bioremediators*, pp. 3–27. Springer, Heidelberg.

Sheng, G.P., Xu, J., Luo, H.W., Li, W.W., Li, W.H., Yu, H.Q., Xie, Z., Wei, S.Q., and Hu, F.C. 2013. Thermodynamic analysis on the binding of heavy metals onto extracellular polymeric substances (EPS) of activated sludge. *Water Research* 47(2): 607–614.

Shukla, O.P., Rai, U.N., and Dubey, S. 2009. Involvement and interaction of microbial communities in the transformation and stabilization of chromium during the composting of tannery effluent treated biomass of *Vallisneria spiralis L. Bioresource Technology* 100: 2198–2203.

Singh, J., and Kalamdhad, A.S. 2012. Reduction of heavy metals during composting-a review. *International Journal of Environmental Protection* 2(9): 36–43.

Singhal, V., and Rathore, V.S. 2001. Effects of Zn^{2+} and Cu^{2+} on growth, lignin degradation and ligninolytic enzymes in *Phanerochaete chrysosporium. World Journal of Microbiology and Biotechnology* 17: 235–240.

Tang, L., Zeng, G.M., Shen, G.L., Zhang, Y., Huang, G.H., and Li, J.B. 2006. Simultaneous amperometric determination of lignin peroxidase and manganese peroxidase activities in compost bioremediation using artificial neural networks. *Analytica Chimica Acta* 579: 109–116.

Tuomela, M., Vikman, M., Hatakka, A., and Itävaara, M., 2000. Biodegradation of lignin in acompost environment: A review. *Bioresource Technology* 72: 169–183.

Varma, V.S., and Kalamdhad, A.S. 2014. Stability and microbial community analysis during rotary drum composting of vegetable waste. *International Journal of Recycling of Organic Waste in Agriculture* 3(52): 1–9.

Vishan, I., Kanekar, H., and Kalamdhad, A. 2014. Microbial population, stability and maturity analysis of rotary drum composting of water hyacinth. *Biologia* 69(10): 1303–1313.

Vishan, I., Laha, A., and Kalamdhad, A. 2017. Biosorption of Pb(II) by Bacillus badius AK strain originating from rotary drum compost of water hyacinth. *Water Science & Technology* 75(5): 1071–1083.

Wei, Y.J., and Liu, Y.S. 2005. Effects of sewage sludge compost application on crops and cropland in a 3-year field study. *Chemosphere* 59: 1257–1265.

Xu, P., Liu, L., Zeng, G., Huang, D., Lai, C., Zhao, M., Huang, C., Li, N., Wei, Z., Wu, H., Zhang, C., Lai, M., and He, Y. 2014. Heavy metal-induced glutathione accumulation and itsrole in heavy metal detoxification in *Phanerochaete chrysosporium. Applied Microbiology and Biotechnology* 98: 6409–6418.

Yang, Y., Mathieu, J.M., Chattopadhyay, S., Miller, J.T., Wu, T., Shibata, T., Guo, W., and Alvarez, P.J. 2012. Defense mechanisms of Pseudomonas aeruginosa PAO1 against quantum dots and their released heavy metals. *ACS Nano* 6(7): 6091–6098.

Yetis, U.A., Dolek, A., Dilek, F.B., and Ozcengiz, G. 2000. The removal of Pb(II) by *Phanerochaete chrysosporium. Water Research* 34: 4090–4100.

Zeng, G., Huang, D., Huang, G., Hu, T., Jiang, X., Feng, C., Chen, Y., Tang, L., and Liu, H. 2007. Composting of lead-contaminated solid waste with inocula of white-rot fungus. *Bioresource Technology* 98: 320–326.

Zeng, G.M., H.L., Huang, D.L., Huang, X.Z., Yuan, R.Q., Jiang, M., Yu, H.Y., Yu, J.C., Zhang, R.Y., Wang, X.L., and Liu, Y. 2009. Effect of inoculating white-rot fungus during different phases on the compost maturity of agricultural wastes. *Process Biochemistry* 44: 396–400.

Zeng, G., Li, N., Huang, D., Lai, C., Zhao, M., Huang, C., Wei, Z., Xu, P., Zhang, C., and Cheng, M. 2015. The stability of Pb species during the Pb removal process by growing cells of *Phanerochaete chrysosporium*. *Applied Microbiology and Biotechnology* 99: 3685–3693.

Zhang, C., Xu, Y., Zhao, M., Rong, H., and Zhang, K. 2018. Influence of inoculating white-rot fungi on organic matter transformations and mobility of heavy metals in sewage sludge based composting. *Journal of Hazardous Materials* 344: 163–168.

Zhao, M., Xu, Y., Zhang, C., Rong, H., and Zeng, G. 2016. New trends in removing heavy metals from wastewater. *Applied Microbiology and Biotechnology* 100: 1–10.

Zorpas, A.A., Arapoglou, D., and Panagiotis, K. 2003. Waste paper and clinoptilolite as a bulking material with dewatered anaerobically stabilized primary sewage sludge (DASPSS) for compost production. *Waste Management* 23(1): 27–35.

9 Leachability of Heavy Metals during the Composting Process

9.1 LEACHABILITY OF HEAVY METALS

The leaching of heavy metals may be defined as the ratio of the quantity of a heavy metal extracted out through toxicity characteristics leaching procedure (TCLP) to its total concentration. This experiment is usually applied to evaluate the leachability of heavy metals in the compost, soil, or sediment (Singh and Kalamdhad, 2013a,b,c). The TCLP experiment is designed to evaluate the agility of organic and inorganic analytes present in the liquid and solid wastes.

According to US EPA (1992), the metal concentration in an extract of the TCLP experiment is higher than that of regulatory limits of heavy metals and specifies the hazardous nature of waste. TCLP is generally applied to evaluate the appropriateness of compost for land application or may be reflected a hazardous waste. This method is considered to simulate the leaching ability of the compost material after its application to soil. The regulatory limits for heavy metals which are leached out under normal environmental conditions are made on the basis of protecting groundwater from metal contamination (Chiroma et al., 2012). For US EPA's (1992) given threshold limit for heavy-metal contamination in mg/kg, (see Table 9.1).

It is very important to examine the leaching potential of heavy metals to govern the appropriateness of unpolluted compost or sludge to be applied in agricultural lands (Pathak et al., 2009). Wang et al. (2010) reported that the leaching fraction of Cu was increased during the thermophilic phase of the composting process, whereas this fraction was decreased at the end of the composting process. The leaching fraction of Cu was found to be approximately 20% (of the total fraction) during the composting process (Villasenor et al., 2011). Hargreaves et al. (2008) reported that the leachable fraction of Ni was increased in the final composting process of municipal solid waste (MSW). Villasenor et al. (2011) reported that the leaching concentrations of heavy metals with clinoptilolite were decreased while increasing the leaching fraction of pH during the composing process. Ciba et al. (1999) reported that the leaching fraction of Zn was found to be higher at pH 2.5 than at pH 4.5. These results confirmed that released metals are the carbonate-bound fraction. Chiang et al. (2007) reported that the leaching concentrations of Zn, Cu, and Ni decreased with increasing composting maturity time. This study also reported that leaching concentration of Zn was found to be higher than that of the other metals. Subsequently, most soluble Zn was leached out during the composting of sewage sludge. However, the leaching fraction of heavy metals with additives was decreased while increasing the

TABLE 9.1
Threshold Limits for Leachable Heavy Metals for Compost/Biosolid

Metals/Metalloids	Threshold limit (mg/kg)
Barium	2000
Arsenic, chromium, silver, and lead	100 for each metal
Selenium and cadmium	20 for each metals
Mercury	4

pH of the composting mixture. In addition, additives had larger surface sites and ion exchange capacity for the immobilization of the heavy metals.

Wang et al. (2008) studied the TCLP-extractable metal contents in the sewage sludge composting with sodium sulfide and lime (SSL) and reported that the TCLP-extractable concentration of Zn, Cu, and Ni decreased with increasing SSL by addition of SSL and reduce availability of metals due to the formation of less soluble metal sulfides and hydroxides. SSL addition of 3% (w/w, dw) in the composing biomass significantly reduced the TCLP-extractable metals as compared to 5% (w/w, dw) of SSL addition. Komilis et al. (2011) reported that the concentrations of Cu, Fe, Zn, and Mn were higher than that of regulatory limits. The concentrations of almost all these metals were found to be below approximately 10 mg L^{-1} (except Fe) during the composting process. The concentrations of Fe decreased over the composting period, whereas the concentrations of Zn and Mn were found to be dominant during the composting process. The metal concentrations increased with the increasing battery content within the bioreactors. The concentrations of Co and Cr were found below the limit of detection during the composting process. Komilis et al. (2011) concluded that the leachable metal concentrations compared with the leaching criteria set by European Decision 2003/33/EC (EC, 2003) and found that the MSW compost could be directed to a nonhazardous.

Singh and Kalamdhad (2013b) reported that leachability of heavy metals was reduced significantly in the agitated pile composting of water hyacinth. Table 9.2 illustrates the changes in the leachable fraction of heavy metals in the composting of water hyacinth. The concentration of Cu was reduced from 5.2% to 0.7% of total Cu. The fact that the leachability of Cu and Zn was reduced might be due to formation of Cu complex with organic matter or humic-like substances at the end of the composting process. The concentration of Mn was reduced from 5.0% to 1.2% of total Mn at the end of the composting period. The concentrations of Fe and Zn were increased at the end of the composting period. Leachability of Ni, Pb, and Cd contents were not detected during the agitated pile composting process of water hyacinth (Singh and Kalamdhad, 2013a). The leachable concentration of Cr was reduced from 1.2% to 0.8% of total Cr at the end of the composting period. Consequently, it forms the most stable complex with humic substances (Qiao and Ho, 1997). According to Singh and Kalamdhad (2013a), the order of leachable heavy-metal contents in the composted water hyacinth was Mn > Zn > Fe=Cr > Cu. Singh and Kalamdhad (2013a)

TABLE 9.2
Changes in Leaching Concentration of Heavy Metals during the Water Hyacinth Composting

Composting Methods	Days	Leachable Concentration of Heavy Metals (mg/kg)							
		Zn	Cu	Mn	Fe	Ni	Pb	Cd	Cr
Agitated pile	0	3.1	1.6	28.5	15.6	ND	ND	ND	3.2
	30	4.8	0.7	13.7	26.5	ND	ND	ND	1.1
Rotary drum/reactor	0	22.2	6.6	205.1	121.7	8.6	53	1.9	6.1
	20	21.3	4.7	175.5	63.3	3.6	17	1.0	3.65

Note: Composition of compost materials water hyacinth (90 kg), sawdust (15 kg), and cattle manure (45 kg).
ND, not detected.

stated that concentrations of these metals were reduced at the end of the composting period (Table 9.2). The concentration of Cu was reduced from 12.2% to 5.3% of the total Cu during the composting process. The leachable concentration of Cr was reduced from 10.3% to 4.8% of the total Cr in the composting process. Chiang et al. (2007) reported the reduction of the leachable fraction of Ni, Zn, and Cu in the composting process of sewage sludge.

Paradelo et al. (2011) reported that the total concentrations of <2% were extracted with the TCLP test, indicating a low concentration of heavy metals involved in potential environmental risk. The order of the extractability of metals was as follows: Zn > Cu > Pb > Ni; however, Cr was not detected in the MSW compost. Chiang et al. (2007) also detected a similar extractability series for sewage sludge compost as follows: Zn > Ni > Cu > Pb=Cr. Paradelo et al. (2011) suggested that the legal limits for possibly toxic trace elements in compost is not only regulated by their total concentration but also by their availability. Paradelo et al. (2011) also suggested that Cu, Pb, and Zn should consider for legal limits, these metals generally not included in legal limit of metals which are usually present in the highest mobile forms. Singh et al. (2014) reported that the leachability of selected heavy metals decreased during the composting of green phumdi biomass. The order of heavy metals extracted by TCLP was as follows: >54.65% for Zn > 54.07% for Cd > 51.78% for Cu > 51.47% for Ni > 47.47% for Cr > 46.51% for Fe > 43.62% for Mn > 28.0% for Pb. Singh et al. (2015) studied the reduction of the leaching potential of heavy metals in agitated pile composting of *Salvinia natans* weed of Loktak lake, Manipur, India. This study reported that the leachability of heavy metals was decreased about 21.2%, 35.0% 18.9%, 34.7%, 30.1%, 34.5%, 50.5%, and 60.0% for Zn, Cu, Mn, Fe, Ni, Pb, Cd, and Cr, respectively, during the composting of *S. natans*. Therefore, the results of the TCLP test confirmed that the concentrations of all selected heavy metals were found to be under the threshold limits of compost which can be used as soil amendment (Singh et al., 2015).

Singh et al. (2016) studied the changes in the leachability of Zn, Cu, Mn, Fe, Ni, Cd, Pb, and Cr during the composting of water fern. This study also reported that leachable fractions of all potential toxic metals were decreased in the composting process. Leachability of heavy metals was reduced approximately 44% for Zn, 60% for Cu, 53% for Mn, 53% for Fe, 49% for Ni, 56% for Pb, and 65% for Cr. This reduction can be attributed to the very good degradation of composting biomass by microbes and conversion into humus-like substances, which formed of complexes with the leachable fraction of heavy metals (Singh et al., 2016). Vishan et al. (2017) studied the changes of leachability of Pb, Cd, Zn, and Ni during the 20 days of the composting period of water hyacinth. This study also reported that the concentration of Pb decreased from the initial stage to the final stage of the composting process. However, the concentrations of Cd and Ni were also reduced from the initial stage to the final stage of the composting process. Similarly, the concentration of Zn also increased from the initial stage to the final stage of the composting process. This study concluded that the reduction of heavy-metal leachability was very efficient for the degradation of the organic matter by the microorganisms during the composting process (Vishan et al., 2017).

Hazarika et al. (2017) reported that the order of concentration of leachability of heavy metal during the composting of paper mill sludge was as follows: Zn > Cd > Mn > Hg > Fe > Ni > Cr > Pb > Cu. The leachable concentrations of Ni, Pb, Cr, Zn, Mn, Cr, and Cu were reduced significantly in all trials at the end of the composting process, whereas the leachable concentrations of these metals were increased in trial 4 at the end of composting process. A leaching behavior of Cd and Fe was slightly different from the Ni, Pb, Cr, Zn, and Cu. The concentration of Hg was not detected in trial 5, whereas this concentration was increased from 5.8% to 6.3% in trial 3 at the end of the composting process of paper mill sludge.

9.2 EFFECTS OF CHEMICAL ADDITION ON LEACHABILITY OF HEAVY METALS IN COMPOSTING

9.2.1 Effects of Waste Lime on Leaching of Heavy Metals

Singh and Kalamdhad (2013c) and Singh and Kalamdhad (2014a) studied the effects of waste lime on the leachability of heavy metals during the agitated pile and rotary drum composting of water hyacinth. Table 9.3 displays that the leaching fractions of Cd, Cr, Cu, Fe, Mn, Ni, and Zn decreased during the agitated pile and rotary drum composting

TABLE 9.3
Effects of Waste Lime on Leaching Concentration of Heavy Metals in Agitated Pile and Rotary Drum Composting of Water Hyacinth

		Leaching Concentration of Heavy Metals (mg/kg)							
Composting Methods	Days	Zn	Cu	Mn	Fe	Ni	Pb	Cd	Cr
Agitated pile composting	0	64.7	11.8	258.4	124.1	1.90	14.6	1.12	11.2
	30	34.4	6.8	205.0	64.6	1.53	10.7	0.26	7.1
Rotary drum composting	0	61.2	5.8	220.9	88.4	16.97	11.685	2.05	7.94
	20	27.45	2.2	144.75	35.4	8.19	4.92	0.79	3.38

of water hyacinth by lime addition. The concentrations of these metals in leaching extract were found to be lower than that of threshold limits given by US EPA (1992). This compost is not considered hazardous for its applications in agricultural lands. The leaching concentrations of Cu and Pb increased in the lime-treated compost of the pile composting process, whereas in the rotary drum composting of water hyacinth only Cu was reduced significantly. The sequence of metals extracted with TCLP reagents in the pile composting process of water hyacinth was as follows: Mn > Zn > Cu > Cr > Pb > Ni > Fe > Cd, whereas in the rotary drum composting process this order was as follows: Mn > Zn > Cu > Cr > Ni > Cd > Pb > Fe. Results of the TCLP experiment confirmed that composts made from water hyacinth through both the agitated pile and rotary drum composting of water hyacinth with lime were appropriate for the land applications. The decrease in the leaching concentration of heavy metals during composting process can be attributed to the higher degradation of organic materials present in the composting mixture following the development of humic-like substances, which had a capability to form metal complex (Wong and Fang, 2000).

Changing pH during the water hyacinth composting process also influences the solubility of the hydroxides and carbonates form of metals. The acidic condition of composting biomass enhanced the solubility of heavy metals causing larger leaching fraction of heavy metals (Qiao and Ho, 1997). Pardo et al. (2011) reported that carbon dioxide is formed during degradation of the organic materials of composting mixture, which involved in maintaining equilibrium of the carbonate/bicarbonate, therefore metal precipitation is influenced by the development of carbonate, oxides or hydroxides of metals; consequently , decreased in leachable fraction of metals (Pardo et al., 2011). Chiang et al. (2007) described that leaching fraction of Pb, Cd, and Cr were not detected in the composting of sewage sludge with lime.

9.2.2 Effects of Natural Zeolite on the Leachability of Heavy Metals

Singh et al. (2013) and Singh and Kalamdhad (2014b) studied the effect of natural zeolite on the extractability of heavy metals through the TCLP experiment during the agitated pile and rotary drum composting of water hyacinth. Table 9.4 illustrates the

TABLE 9.4
Effects of Natural Zeolite on Heavy-Metal Extraction through TCLP Experiment during Agitated Pile Composting and Rotary Drum Composting of Water Hyacinth

		Leaching Concentration of Heavy Metals (mg/kg)							
Composting Methods	**Days**	**Zn**	**Cu**	**Mn**	**Fe**	**Ni**	**Pb**	**Cd**	**Cr**
Agitated pile composting	0	42.53	12.0	259.50	31.02	13.42	37.0	1.26	5.70
	30	18.86	4.38	216.90	16.44	7.60	27.0	0.43	4.08
Rotary drum composting	0	38.15	7.31	180.40	27.10	5.50	22.05	2.95	4.30
	20	21.28	3.60	130.0	23.40	2.10	12.65	1.22	1.41

difference of the leaching fraction of Mn, Fe, Ni, Pb, Cd, Zn, Cu, and Cr in the composting process. The leaching fractions of heavy metals were decreased in the range of 25.7%–67.4% for Zn, 31.6%–73.2% for Cu, 18.2%–52.7% for Mn, 35.4%–67.0% for Fe, 44.3%–67.9% for Ni, 32.8%–72.4% for Pb, 46.7%–71.0% for Cd, and 42.2%–72.6% for Cr in control and all zeolite treatments during the water hyacinth composting.

The leachability of Zn, Mn, Fe, and Ni increased approximately 5%, whereas the leachability of Cd and Cr decreased up to 10% in zeolite in the composting process. The addition of natural zeolite caused a significant increase in the pH of the initial composting mixture as compared with control, leading to a decrease in the leachable fractions of heavy metals (Su and Wong, 2003). The decrease in the leaching fraction of metals during the composting process can be attributed as at the end of composting process humic substances are developed which have high affinity to metal and form metal humus complex (Wong and Fang, 2000; Villasenor et al., 2011). The leaching fractions of Cu and Pb increased significantly in zeolite treatments as compared with control. It has been observed that higher percentage of zeolite addition reduced the leaching fraction of metals. Leachability of heavy metals decreased up to 15% in zeolite treatment as compared with control and 5% and 10% in zeolite treatments during the rotary drum composting of water hyacinth.

9.3 CONCLUSION

The heavy-metal concentrations decreased in the composting of water hyacinth, sewage sludge, phumdi biomass, and municipal solid compost *Salvinia natans* weed and found that metal concentrations were under the threshold limits for compost used in agricultural lands. Leaching quantity of heavy metals was measured under the threshold limits in the final composting of water hyacinth. The addition of waste lime was highly effective for decreasing leachability of heavy metals both during agitated pile composting and rotary drum composting of water hyacinth. The maximum reduction in leachability of heavy metals was achieved in lime treatments (1% and 2%) which specified that the optimum amount of lime addition can enhance organic matter degradation followed by humification process in a rotary drum composting of water hyacinth, subsequently toxicity of the metals was reduced significantly. The addition of lime in the composting mixture may be beneficial not only for reducing leachability of metals but also for waste management practices. Addition of SSL reduced the heavy-metal availability during the composting of sewage sludge. The maximum reduction of leachability of heavy metals was found to be 5% and 10% with zeolite addition this may be due to the ion exchange capacity of zeolite with metals.

REFERENCES

Chiang, K.Y., Huang, H.J., and Chang, C.N. 2007. Enhancement of heavy metal stabilization by different amendments during sewage sludge composting process. *Journal of Environmental Engineering Management* 17(4): 249–256.

Chiroma, T.M., Ebewele, R.O., and Hymore, F.K. 2012. Levels of heavy metals (Cu, Zn, Pb, Fe and Cr) in Bushgreen and Roselle irrigated with treated and untreated urban sewage water. *International Research Journal of Environmental Sciences* 1(4): 50–55.

Ciba, J., Korolewicz, T., and Turek, M. 1999. The occurrence of metals in composted municipal wastes and their removal. *Water, Air, & Soil Pollution* 111: 159–170.

European Council. 2003. Decision 2003/33/EC on the establishment of criteria and procedures for the acceptance of waste at landfills pursuant to article 16 of and Annex II to Directive 1999/31/EC. *Official Journal of the European Union* L11: 27–49.

Hargreaves, J.C., Adl, M.S., and Warman, P.R. 2008. A review of the use of composted municipal solid waste in agriculture. *Agriculture, Ecosystems and Environment* 123: 1–14.

Hazarika, J., Ghosh, U., Kalamdhad, A.S., Khwairakpam, M., and Singh, J. 2017. Transformation of elemental toxic metals into immobile fractions in paper mill sludge through rotary drum composting. *Ecological Engineering* 101: 185–192.

Komilis, D., Bandi, D., Kakaronis, G., and Zouppouris, G. 2011. The influence of spent household batteries to the organic fraction of municipal solid wastes during composting. *Science of the Total Environment* 409(13): 2555–2566.

Paradelo, R., Villada, A., Devesa-Rey, R., Moldes, A.B., Dominguez, M., Patino, J., and Barral, M.T. 2011. Distribution and availability of trace elements in municipal solid waste composts. *Journal of Environmental Monitoring* 13: 201–211.

Pardo, T., Clemente, R., and Bernal, M.P. 2011. Effects of compost, pig slurry and lime on trace element solubility and toxicity in two soils differently affected by mining activities. *Chemosphere* 84: 642–650.

Pathak, A., Dastida, M.G., and Sreekrishnan, T.R. 2009. Bioleaching of heavy metals from sewage sludge: A review. *Journal of Environmental Management* 90: 2343–2353.

Qiao, L., and Ho, G. 1997. The effects of clay amendment and composting on metal speciation in digested sludge. *Water Research* 31(5): 951–964.

Singh, J., and Kalamdhad, A.S. 2013a. Assessment of bioavailability and leachability of heavy metals during rotary drum composting of green waste (water hyacinth). *Ecological Engineering* 52: 59–69.

Singh, J., and Kalamdhad, A.S. 2013b. Bioavailability and leachability of heavy metals during water hyacinth composting. *Chemical Speciation and Bioavailability* 25(1): 1–14.

Singh, J., and Kalamdhad, A.S. 2013c. Effects of lime on bioavailability and leachability of heavy metals during agitated pile composting of water hyacinth. *Bioresource Technology* 138: 148–155.

Singh, J., and Kalamdhad, A.S. 2014a. Effects of carbide sludge (lime) on bioavailability and leachability of heavy metals during rotary drum composting of water hyacinth. *Chemical Speciation and Bioavailability* 2: 1–9.

Singh, J., and Kalamdhad, A.S. 2014b. Uptake of heavy metals by natural zeolite during agitated pile composting of water hyacinth composting. *International Journal of Environmental Science* 5(1): 1–15.

Singh, J., Prasad, R., and Kalamdhad, A.S. 2013. Effect of natural zeolite on bioavailability and leachability of heavy metals during rotary drum composting of green waste. *Research Journal of Chemical and Environmental Sciences* 17: 26–34.

Singh, W.R., Pankaj, S., Singh, J., and Kalamdhad, A.S. 2014. Evaluation of bioavailability of heavy metals and nutrients during agitated pile composting of green Phumdi. *Research Journal of Chemical and Environmental Sciences* 18: 1–8.

Singh, J., Kalamdhad, A.S., and Lee, B.K. 2015. Reduction of eco-toxicity risk of heavy metals in the rotary drum composting of water hyacinth: Waste lime application and mechanisms. *Environmental Engineering Research* 20(3): 212–222.

Singh, W.R., Kalamdhad, and A.S., Singh, J. 2016. The preferential composting of water fern and a reduction of the mobility of potential toxic elements in a rotary drum reactor. *Process Safety and Environmental Protection* 102: 485–494.

Su, D.C., and Wong, J.W.C. 2003. Chemical speciation and phytoavailability of Zn, Cu, Ni and Cd in soil amended with fly ash-stabilized sewage sludge. *Environment International* 29: 895–900.

US Environmental Protection Agency. 1992. Method 1311—Toxicity characteristic leaching procedure (TCLP), 35 p.

Villasenor, J., Rodriguez, L., and Fernandez, F.J. 2011. Composting domestic sewage sludge with natural zeolites in a rotary drum reactor. *Bioresource Technology* 102(2): 1447–1454.

Vishan, I., Sivaprakasam, S., and Kalamdhad, A. 2017. Isolation and identification of bacteria from rotary drum compost of water hyacinth. *International Journal of Recycling of Organic Waste in Agriculture* 6(3): 245–253.

Wang, X., Chen, L., Xia, S., and Zhao, J. 2008. Changes of Cu, Zn, and Ni chemical speciation in sewage sludge co-composted with sodium sulfide and lime. *Journal of Environmental Sciences* 20: 156–160.

Wang, X.D., Chen, X.N., Ali, A.S., Liu, S., and Lu, L.L. 2010. Dynamics of humic substance-complexed copper and copper leaching during composting of chicken manure. *Pedosphere* 20(2): 245–251.

Wong, J.W.C., and Fang, M. 2000. Effects of lime addition on sewage sludge composting process. *Water Research* 34(15): 3691–3698.

Index